LISA RANDALL

Lisa Randall is one of the world's leading theoretical physicists and the Frank B. Baird, Jr. Professor of Science at Harvard University. Her research involves connecting theoretical insights to puzzles in our current understanding of the properties and interactions of matter. Her past work has included a model about the structure of an extra dimension of space, and she is currently working on the nature of dark matter. Randall has received numerous awards and honours and is a member of the National Academy of Sciences, the American Philosophical Society, the American Academy of Arts and Sciences, an Honorary Member of the Royal Irish Academy and an Honorary Fellow of the Institute of Physics. She is the bestselling author of several acclaimed books including *Knocking on Heaven's Door* and *Higgs Discovery*.

ALSO BY LISA RANDALL

Warped Passages

Knocking on Heaven's Door

Higgs Discovery

LISA RANDALL

Dark Matter and the Dinosaurs

The Astounding Interconnectedness of the Universe

VINTAGE

1 3 5 7 9 10 8 6 4 2

Vintage
20 Vauxhall Bridge Road,
London SW1V 2SA

Vintage is part of the Penguin Random House
group of companies whose addresses can be found at
global.penguinrandomhouse.com

First published in Vintage in 2017
First published in hardback by The Bodley Head in 2015

penguin.co.uk/vintage

A CIP catalogue record for this book is
available from the British Library

ISBN 9780099593560

Printed and bound by Clays Ltd, St Ives Plc

CONTENTS

PART III:
DECIPHERING DARK MATTER'S IDENTITY

INTRODUCTION

"Dark matter" and "dinosaurs" are words you rarely hear together except perhaps in the playground, a fantasy gaming club, or some not-yet-released Spielberg movie. Dark matter is the elusive stuff in the Universe that interacts through gravity like ordinary matter, but that doesn't emit or absorb light. Astronomers detect its gravitational influence, but they literally don't see it. Dinosaurs, on the other hand . . . I doubt I need to explain dinosaurs. They were the dominant terrestrial vertebrates from 231 to 66 million years ago.

Though both dark matter and dinosaurs are independently fascinating, you might reasonably assume that this unseen physical substance and this popular biological icon are entirely unrelated. And this might well be the case. But the Universe is by definition a single entity and in principle its components interact. This book explores a speculative scenario in which my collaborators and I suggest that dark matter might ultimately (and indirectly) have been responsible for the extinction of the dinosaur.

Paleontologists, geologists, and physicists have shown that 66 million years ago, an object at least ten kilometers wide plummeted to Earth from space and destroyed the terrestrial dinosaurs, along with three-quarters of the other species on the planet. The object might have been a comet from the outer reaches of the Solar System, but no one knows why this comet was perturbed from its weakly bound, but stable, orbit.

Our proposal is that during the Sun's passage through the mid-plane of the Milky Way—the stripe of stars and bright dust that you can observe in a clear night sky—the Solar System encountered a disk of dark matter that dislodged the distant object, thereby precipitating this cataclysmic impact. In our galactic vicinity, the bulk of the dark matter surrounds us in an enormous smooth and diffuse spherical halo.

The type of dark matter that triggered the dinosaurs' demise would be distributed very differently from most of the elusive dark matter in the Universe. The additional type of dark matter would leave the halo intact, but its very different interactions would make it condense into a disk—right in the middle of the Milky Way plane. This thin region could be so dense that when the Solar System passes through it, as the Sun oscillates up and down during its orbit through our galaxy, the disk's gravitational influence would be unusually strong. Its gravitational pull could be powerful enough to dislodge comets at the outer edge of the Solar System, where the Sun's competing pull would be too weak to rein them back in. The errant comets would then be ejected from the Solar System or—more momentously—be redirected to hurtle toward the inner Solar System, where they might have the potential to strike the Earth.

I'll tell you right up front that I don't yet know if this idea is correct. It's only an unexpected type of dark matter that would yield measurable influences on living beings (well, technically no longer living). This book is the story of our unconventional proposal about just such surprisingly influential dark matter.

But these speculative ideas—as provocative as they might be—are not this book's primary focus. At least as important to its content as the story of the dinosaur-destroying comet are the context and the science that embrace it, which include the far better established frameworks of cosmology and the science of the Solar System. I feel very fortunate that the topics I study frequently guide my research toward big questions such as what stuff is made of, the nature of space and time, and how everything in the Universe evolved to the world we see today. In this book, I hope to share a lot of this too.

In the research that I will describe, my studies led me down a path where I started thinking more broadly about cosmology, astrophysics, geology, and even biology. The focus was still on fundamental physics. But having done more conventional particle physics all my life—the study of the building blocks of familiar matter such as the paper or screen on which you're reading this—I've found it refreshing to probe into what is known—and what soon will be known—about the dark world too, as well as the implications of basic physical processes for the Solar System and for the Earth.

Dark Matter and the Dinosaurs explains our current knowledge about the Universe, the Milky Way, the Solar System, as well as what makes for a habitable zone and life on Earth. I'll discuss dark matter and the cosmos, but I will also delve into comets, asteroids, and the emergence and extinction of life, with special focus on the object that fell to Earth to kill off the terrestrial dinosaurs—and a lot of the rest of life here. I wanted this book to convey the many incredible connections that got us here so we can more meaningfully understand what is happening now. When we think about our planet today, we might also want to better understand the context in which it developed.

When I started concentrating on the concepts underlying the ideas in this book, I was awe-struck and enchanted not only by our current knowledge of our environment—local, solar, galactic, and universal, but also by how much we ultimately hope to understand, from our random tiny perch here on Earth. I also was overwhelmed by the many connections among the phenomena that ultimately allow us to exist. To be clear, mine is not a religious viewpoint. I don't feel the need to assign a purpose or meaning. Yet I can't help but feel the emotions we tend to call religious as we come to understand the immensity of the universe, our past, and how it all fits together. It offers anyone some perspective when dealing with the foolishness of everyday life.

This newer research actually has made me look differently at the world and the many pieces of the Universe that created the Earth—and us. Growing up in Queens I saw the impressive buildings of New York City, but not so much of nature. What little nature I did see was cultivated into parks or lawns—retaining little of the form it took before humans arrived. Yet when you walk on a beach, you are walking on ground up creatures—or at least their protective coverings. The limestone cliffs you might see on a beach or in the countryside are composed of previously living creatures too, from millions of years in the past. Mountains arose from tectonic plates that collided, and the molten magma that drives these movements is the result of radioactive material buried near the core of the Earth. Our energy came from the Sun's nuclear processes—though it has been transformed and stored in different ways since those initial nuclear reactions occurred. Many of the resources we use are heavier elements that came from outer space, which were deposited on the Earth's surface by asteroids or comets. Some amino acids were deposited by meteoroids too—perhaps bringing life—or the seeds of life—to Earth. And before any of this happened, dark matter collapsed into clumps whose gravity attracted more matter—which eventually turned into galaxies, galaxy clusters, and stars like our Sun. Ordinary matter—important as it is to us— does not tell the whole story.

Although we might experience the illusion of a self-contained environment, every day at sunrise and every night when the Moon and the far more distant stars come into view, we are reminded that our planet is not alone. Stars and nebulae are further evidence that we exist in a galaxy that resides within a far larger Universe. We orbit within a Solar System where the seasons remind us further of our orientation and placement within it. Our very measurement of time in terms of days and years signifies the relevance of our surroundings.

• • •

Four inspiring lessons that I wanted to share stand out to me from the research and readings that led to this book. Close to my heart is the satisfaction of understanding how the pieces of the Universe connect in so many remarkable ways. The big lesson at the most fundamental level is that the physics of elementary particles, the physics of the cosmos, and the biology of life itself all connect—not in some New-Age sense, but in remarkable ways that are well worth understanding.

Stuff from outer space hits the Earth all the time. Yet the Earth has a love-hate relationship with its environment. The planet benefits from some of what's around us, but much of it can be lethal. The position of our planet allows for the right temperature, the outer planets divert most incoming asteroids and comets before they strike the Earth, the distance between the Moon and the Earth stabilizes our orbit sufficiently to prevent massive temperature fluctuations, and the outer Solar System shields us from dangerous cosmic rays. Meteoroids hitting the Earth might have deposited resources critical to life, but they also affected the trajectory of life on the planet in more detrimental ways. At least one such object led to a devastating extinction 66 million years ago. Though it wiped out the land-dwelling dinosaurs, it also paved the way for the existence of larger mammals, including ourselves.

The second point—also impressive—is how recent are so many of the scientific developments that I will discuss. Perhaps people can make the following statement at any point in human history, but that does not diminish its validity: we have advanced our knowledge tremendously in the last [here insert a context-dependent number] years. For the research I will describe, that number is less than fifty. As I was doing my own research, and reading about others', I was constantly struck by how new and deeply revolutionary so many recent discoveries have been. Human ingenuity and stubbornness have consistently emerged as scientists have tried to reconcile themselves to the often surprising and always entertaining and sometimes scary things we learned about the world. The science this book presents is part of a larger history—13.8 or 4.6 billion years according to whether you focus on the Universe or the Solar System. However, the history of human beings' unraveling these ideas is little more than a century old.

The dinosaurs went extinct 66 million years ago, but paleontologists and geologists deduced the nature of that extinction only in the 1970s and 1980s. Once the relevant ideas had been introduced, it was a matter of decades before a community of scientists more fully evaluated them. And the timing was not entirely coincidental. The extinction's connection to an extraterrestrial object became more credible once astronauts had landed on the Moon and seen craters up close—presenting them with detailed evidence of the dynamical nature of the Solar System.

In the last fifty years, significant advances in particle physics and cosmology have taught us about the Standard Model, which describes the basic elements of matter as we understand them today. The amount of dark matter and dark energy in the Universe too was pinned down only in the last decades of the twentieth century. Our knowledge of the Solar System also changed during the same time frame. And only in the 1990s did scientists discover the Kuiper belt objects in Pluto's vicinity, demonstrating that Pluto is not orbiting alone. The number of planets was reduced—but only because the science you might have learned in grade school is now richer and more complex.

The third major lesson centers on the rate of change. Natural selection permits adaptation when species have time to evolve. But that adaptation won't encompass radical changes. It is far too slow. The dinosaurs weren't in a position to prepare for a 10-kilometer-wide meteoroid hitting the Earth. They couldn't adapt. Those stuck on land, who were too big to bury themselves, had nowhere viable to go.

As new ideas or technologies emerge, debates over catastrophic versus gradual change have also played a big role. Key to understanding most new developments—scientific or otherwise—is the pace of the processes they describe. I frequently hear people suggest that certain developments, such as studies in genetics or advances deriving from the Internet, are unprecedentedly dramatic. But this is not entirely true. The improved understanding of disease or of the circulatory system, which dates back hundreds of years, brought about changes at least as profound as genetics does today. The introduction of written language, and later of the printing press, influenced the ways people acquired knowledge and how they thought in ways at least as significant as those that the Internet precipitated.

As with these developments, a very important factor for current change is also its rapidity—a topic that can be pertinent not only to scientific processes, but to environmental and sociological changes too. Although death by meteoroid is not likely to be a significant concern for us today, the quickening rates of changes in the environment and in extinctions likely are—and the impact could be comparable in many ways. The perhaps not-so-hidden agenda of this book is to help us better understand the amazing story of how we got here and to encourage us to use that knowledge wisely.

Even so, the fourth important lesson is the remarkable science describing the often hidden elements of our world and its development—and how much about the Universe we can hope to understand. Many people are fascinated by the idea of a multiverse—other universes not within our reach. But at least as fascinating are the many hidden worlds—both biological and physical —that we do have a chance to explore and learn more about. In *Dark Matter and the Dinosaurs,* I hope to convey how inspiring it can be to contemplate what we know—as well as what we might expect or hope to figure out in the future.

• • •

This book begins by explaining cosmology—the science of how the Universe has evolved to its current state. Its first part presents the Big Bang theory, cosmological inflation, and the makeup of the Universe. This section also explains what dark matter is, how we ascertained its existence, and why it is relevant to the Universe's structure.

Dark matter constitutes 85 percent of the matter in the Universe while ordinary matter—such as that contained in stars, gas, and people—constitutes only 15 percent. Yet people are mainly preoccupied with the existence and relevance of ordinary matter—which, to be fair, interacts far more strongly.

However, as with humanity, it doesn't make sense to focus all our attention on the small percentage that is disproportionately influential. The dominant 15 percent of matter that we can see and feel is only part of the story. I will explain dark matter's critical role in the Universe—both for galaxies and for galaxy clusters forming out of the amorphous cosmic plasma in the early Universe—and in maintaining the stability of these structures today.

The second part of the book zooms in on the Solar System. The Solar System alone could of course be the subject of an entire book, if not of an encyclopedia. So I will focus on the constituents that might have concerned the dinosaurs—meteoroids, asteroids, and comets. This part will describe objects that we know have hit the Earth and what we anticipate might hit it in the future, as well as the sparse but not-obviously-dismissible evidence for extinctions or meteoroid strikes that occur at regularly spaced intervals of about 30 million years. This section also discusses life's formation, as well as its destruction—reviewing what is known about the five major mass extinctions, including the devastating event that killed the dinosaurs.

The book's third and final part integrates the ideas from the first two, starting with a discussion of models of dark matter. It explains the more familiar models for what dark matter might be, as well as the newer suggestion for dark matter interactions hinted at above.

At this point, we know only that dark matter and ordinary matter interact via gravity. Gravity's consequences are generally so tiny that we register the influence only of enormous masses—such as that of the Earth and the Sun—and even those are pretty feeble. After all, you can pick up a paper clip with a tiny magnet, successfully competing against the gravitational influence of the entire Earth.

However, dark matter might experience other forces too. Our new model challenges people's assumption—and prejudice—that familiar matter is unique because of the forces—electromagnetism, the weak, and the strong nuclear forces—through which it interacts. These conventional matter forces, which are much stronger than gravity, account for many of the interesting features of our world. But what if some of the dark matter experiences influential non-gravitational interactions too? If true, dark matter forces could lead to dramatic evidence of connections between elementary matter and macroscopic phenomena even deeper than the many we already know to be present.

Although everything in the Universe could in principle interact, most such interactions are far too small to readily register. Only things that affect us in a detectable way can be observed. If you have something exerting and experiencing only tiny effects, it might be right under your nose yet escape your notice. That's presumably why individual dark matter particles—though probably all around us—have so far escaped discovery.

The third part of the book shows how thinking more broadly about dark matter—asking why the dark universe should be so simple when ours is so complicated—led us to consider some novel possibilities. Maybe a portion of the dark matter experiences is own force—dark light if you will. If most dark matter is usually relegated to the relatively uninfluential 85 percent, we could then think of the newly proposed type of dark matter as an upwardly mobile middle class—with interactions mimicking those of familiar matter. The additional interactions would affect the makeup of the galaxy and allow this portion of dark matter to affect the motion of stars and other objects in the domain of ordinary matter.

In the next five years, satellite observations will measure the galaxy's shape, composition, and properties in greater detail than ever before—telling us a great deal about our galactic environment and testing whether or not our conjecture is true. Such observable implications make dark matter and our model legitimate science that is worthy of exploration—even if dark matter is not a building block of you and me. The consequences might include meteoroid impacts—one of which could have been the link between dark matter and the disappearance of the dinosaurs to which the book's title alludes.

The background and concepts that connect these phenomena offers us a capacious, 3-d picture of the Universe. My goal in writing this book is to share these ideas and to encourage you to explore, appreciate, and bolster the remarkable richness of our world.

PART I

THE DEVELOPMENT OF THE UNIVERSE

1

THE CLANDESTINE DARK MATTER SOCIETY

We often fail to notice things that we are not expecting. Meteors flash across the sky on a moonless night, unfamiliar animals shadow us when we hike through the woods, magnificent architectural details surround us as we walk through a town. Yet we often overlook these remarkable sights—even when they are directly in our field of view. Our very bodies host colonies of bacteria. Ten times more bacterial cells than human cells live inside us and help with our survival. Yet we are barely aware of these microscopic creatures that live in us, consume nutrients, and aid our digestive systems. Only when bacteria misbehave and make us ill do most of us even acknowledge their existence.

To view things, you have to look. And you have to know how to look. But at least the phenomena I just mentioned can in principle be seen. Imagine the further challenges in understanding something that you literally cannot see. That would be dark matter, the elusive stuff in the Universe that has only minuscule interactions with the matter we understand. In the chapter that follows, I'll explain the many measurements with which astronomers and physicists have established dark matter's existence. In this one, I'll introduce this elusive matter: what it is, why it might seem so perplexing, and why—from some important perspectives—it is not.

THE UNSEEN IN OUR MIDST

Although the Internet is a single giant network in which billions of people engage online, most of those who are communicating on social networks don't interact directly—or even indirectly—with each other. Participants tend to friend like-minded people, follow others with similar interests, and turn to news sources that represent their own particular worldview. With such restricted interactions, the many people engaged on-line fragment into distinct, non-interacting populations within which they rarely encounter an objectionable point of view. Even people's friends' friends don't generally confront the contradictory opinions of unaffiliated groups, so most of the Internet's participants are oblivious to the existence of unfamiliar communities with different, incompatible ideas.

We're not all so closed to worlds outside our own. But when it comes to dark matter, we're all guilty as charged. Dark matter just isn't part of ordinary matter's social network. It lives in an Internet chat room that we don't yet know how to enter. It's in the same Universe and even occupies the same regions of space as visible matter. But dark matter particles interact only imperceptibly with the ordinary matter that we know. As with Internet communities to which we are oblivious—unless we are told about dark matter, in our daily lives we would be unaware it exists.

Like the bacteria within us, dark matter is one of many other "universes" right under our noses. And like those microscopic creatures, it is all around us too. Dark matter passes right through our bodies—and resides in the outside world as well. Yet we don't notice any of its consequences because it interacts so feebly—so much so that it forms a distinct population. It is a society totally separate from the matter that we know.

But it's an important one. Whereas bacterial cells—though numerous—account for only about one or two percent of our weight, dark matter—though an insignificant fraction of our bodies—accounts for about 85 percent of the matter in the Universe. Every cubic centimeter around you contains about a proton's mass worth of matter. That might sound like a lot or a little depending on how you view it. But it means that if dark matter is composed of particles whose mass is comparable to those we know of and if those particles travel at the velocity we expect based on well-understood dynamics, billions of dark matter particles pass through each of us every second. Yet no one notices that they are there. The effect of even billions of dark matter particles on us is minuscule.

That's because we can't sense dark matter. Dark matter doesn't interact with light—at least to the extent that people have been able to probe so far. Dark matter is not made out of the same material as ordinary matter—it's not composed of atoms or the familiar elementary particles that do interact with light, which is essential to everything we can see. The mystery that my colleagues and I hope to solve is what precisely dark matter is composed of. Does it consist of a new type of particle? If so, what are its properties? Aside from its gravitational interactions, does it have any interactions at all? If we are lucky with current experiments, particles of dark matter might turn out to experience some minuscule electromagnetic interactions that have so far been too small to detect. Dedicated probes are searching—I'll explain how in the third part of the book. But so far dark matter remains invisible. Its effects haven't influenced detectors at their current level of sensitivity.

However, when large amounts of dark matter aggregate into concentrated regions, its net gravitational influence is substantial, leading to measurable influences on stars and on nearby galaxies. Dark matter affects the expansion of the Universe, the path of light rays passing to us from distant objects, the orbits of stars around the centers of galaxies, and many other measurable phenomena in ways that convince us of its existence. We know about dark matter—and indeed it does exist—because of these measurable gravitational effects.

Furthermore, even though it is unseen and unfelt, dark matter played a pivotal role in forming the Universe's structure. Dark matter can be compared to the under-appreciated rank and file of society. Even when invisible to the elite decision makers, the many workers who built pyramids or highways or assembled electronics were crucial to the development of their civilizations. Like other unnoticed populations in our midst, dark matter was essential to our world.

We wouldn't even be around to comment on any of this, let alone put together a coherent picture of the Universe's evolution, if dark matter hadn't been present in the early Universe. Without dark matter, there wouldn't have been enough time to form the structure that we now observe. Clumps of dark matter seeded the Milky Way galaxy—as well as other galaxies and galaxy clusters. Had galaxies not formed, neither would have the stars, nor the Solar System, nor life as we know it. Even today, the collective action of dark matter keeps galaxies and galaxy clusters intact. Dark matter might even be relevant to the trajectory of the Solar System if the dark disk alluded to in the introduction exists.

Yet we don't observe dark matter directly. Scientists have studied many forms of matter but all of them whose composition we know have been observed with some form of light—or more generally, electromagnetic radiation. Electromagnetic radiation appears as light at visible frequencies, but can also appear as radio waves or ultraviolet radiation, for example, when outside the limited range of frequencies that we can see. The effects might be observed under a microscope, with a radar device, or in optical images on a photograph. But electromagnetic influences are always involved. Not all of the interactions are direct—charged elements interact with light most directly. But the elements of the Standard Model of particle physics—the most basic elements of the matter we know about—interact with each other enough that light, if not directly a friend, is at least a friend of a friend of all the forms of matter we can see.

Not only our vision, but our other senses—touch, smell, taste, and sound—rely on atomic interactions, which rely in turn on the interactions of electrically charged particles. Touch too, though for more subtle reasons, relies on electromagnetic vibrations and interactions. Since human senses are all based in electromagnetic interactions of some sort, we can't directly detect dark matter in the usual ways. Although dark matter is all around us, we can't see or feel it. When light shines on dark matter, it doesn't do anything. The light just passes through.

Given that they've never seen (or felt or smelled) it, many people I've spoken to are surprised at the existence of dark matter and find it quite mysterious—or even wonder if it's some sort of mistake. People ask how it can possibly be that most matter—about five times the amount of ordinary matter—cannot be detected with conventional telescopes. Personally, I would expect quite the opposite (though admittedly not everyone sees it this way). It would be even more mysterious to me if the matter we can see with our eyes is all the matter that exists. Why should we have perfect senses that can directly perceive everything? The big lesson of physics over the centuries is how much is hidden from our view. From this perspective, the question is really why the stuff we do know about should constitute as much of the energy density of the Universe as it does.

Dark matter might sound like an exotic suggestion to some, but proposing its existence is far less rash than revising the laws of gravity—as dark matter skeptics might prefer. Dark matter—although indeed unfamiliar—is likely to have a more or less conventional explanation that is completely consistent with all known physical laws. After all, why should all matter that acts in accordance with known laws of gravity behave exactly like familiar matter? To put it succinctly, why should all matter interact with light? Dark matter can simply be matter that has different or no fundamental charges. Without electric charge or interactions with charged particles, dark matter simply can't absorb or emit light.

However, I do have a slight problem with one aspect of dark matter, which is its name. I don't take issue with the "matter" part. Dark matter is indeed a form of matter—meaning that it is stuff that clumps and exerts its own gravitational influence, interacting with gravity like all other matter. Physicists and astronomers detect its presence in diverse ways that rely on this interaction.

It is the "dark" in the name that is unfortunate—both because we see dark things, which absorb light, and because the ominous-sounding label makes it sound more potent and negative than it actually is. Dark matter is not dark—it is transparent. Dark stuff absorbs light. Transparent things, on the other hand, are oblivious to it. Light can hit dark matter, but neither the matter nor the light will change as a result.

At a recent conference bringing together people from many disciplines, I met Massimo, a marketing professional who specializes in branding. When I told him about my research, he looked at me incredulously and asked, "Why is this called dark matter?" His objection was not to the science, but to the name's needlessly negative connotations. It's not exclusively true that every brand associates negative qualities with "dark." The "Dark Knight" was a good guy—if complicated. But compared to its use in *Dark Shadows, His Dark Materials, Transformers: Dark of the Moon,* Darth Vader's "dark side of the Force"—not to mention the hilarious dark song from the Lego movie—the "dark" in "dark matter" is pretty tame. Despite our evident fascination with things dark, dark matter doesn't really live up to the name's reputation.

Dark matter does, however, share one quality with the evil stuff: it is hidden from view. Dark matter is aptly named in the sense that no matter how much you heat it up, it won't emit light. In that sense it is truly dark—not in the sense of being opaque but in the sense of being the opposite of light-emitting or even light-reflecting. No one sees dark matter directly—even with a microscope or a telescope. As with the many malevolent spirits in movies and literature, its invisibility serves as its shield.

Massimo agreed that "transparent matter" would have been a better name—or at least less scary. Although true from a physics perspective, I'm not certain that he's right. "Dark matter," even if not my favorite term, seems to attract a fair amount of attention. Nonetheless, dark matter is neither ominous nor powerful—at least without a huge amount of it. It interacts so feebly with normal matter that it's extremely challenging to find. That's part of what makes it so interesting.

BLACK HOLES AND DARK ENERGY

The name "dark matter" gives rise to other confusions too—even beyond the ominous-sounding implications referred to above. For example, a lot of the people I talk to about my research fail to distinguish dark matter from black holes. To clarify the distinction, I'll take a brief detour to discuss black holes, which are objects that form when too much matter gets within too small a region of space. Nothing—including light—escapes from their powerful gravitational influence.

Black holes and dark matter are no more the same than black ink and film noir. Dark matter doesn't interact with light. Black holes absorb light—and anything else that comes too close. Black holes are black because all light that goes in remains in. It is not radiated and it is not reflected back. Dark matter might have been relevant to the formation of black holes* since any form of matter can collapse to form one. But black holes and dark matter are certainly not the same thing. They should in no way be confused.

One further misunderstanding is triggered by dark matter's infelicitous name. Because another component of the Universe is named "dark energy"—also a problematic choice—people often confuse it too with dark matter. Although also a diversion from our main topic, dark energy is an essential part of cosmology today. So I'll now clarify this other term to ensure that you—my enlightened reader—will always know the difference.

Dark energy is not matter—it is just energy. Dark energy exists even if no actual particle or other form of stuff is around. It permeates the Universe, but doesn't clump like ordinary matter. The density of dark energy is the same everywhere—it can be no denser in one region than another. It is very different from dark matter, which collects into objects and will be denser in some places than in others. Dark matter acts like familiar matter, which gets bound into objects

---

* To be precise, black holes have been proposed as a possible dark matter candidate—a topic we will get to later. Observational constraints and theoretical issues now make this scenario very unlikely.

such as stars, galaxies, and galaxy clusters. The dark energy distribution, on the other hand, is always smooth.

Dark energy also remains constant over time. Unlike matter or radiation, dark energy does not become more dilute when the Universe expands. This is in some respects its defining property. The dark energy density—energy not carried by particles or matter—remains the same over time. For this reason, physicists often refer to this type of energy as a *cosmological constant*.

Early in the Universe's evolution, most of the energy was carried by radiation. But radiation dilutes more quickly than matter so matter took over eventually as the largest energy contribution. Much later in the Universe's evolution, dark energy—which never diluted whereas both radiation and matter did—came to dominate and now constitutes about 70 percent of the Universe's energy density.

Before Einstein proposed his theory of relativity, people thought only about relative energy—the difference in energy between one setup and another. But equipped with Einstein's theory, we learned that the absolute amount of energy is itself meaningful and produces a gravitational force that can contract or expand the Universe. The big mystery about dark energy is not why it exists—quantum mechanics and the theory of gravity suggest it should be present and Einstein's theory tells us it has physical consequences—but why its density is so low. Given it's dominance, this might not seem to be an issue. But although dark energy makes up most of the Universe's energy today, it is only recently—after matter and radiation were diluted enormously by the Universe's expansion—that the influence of dark energy began to compete with that of the other types of energy. Earlier on, the dark energy density was minuscule compared to the other much larger radiation and matter contributions. Without knowing the answer in advance, physicists would have estimated that the dark energy density should be an astounding 120 orders of magnitude bigger. The question of the small size of the cosmological constant has flummoxed physicists for years.

Many astronomers say that we now live in a renaissance era of cosmology, in which theories and observations have advanced to the stage where precisely calibrated tests will help pin down which ideas are realized in the Universe. However, given the dominance of dark energy and dark matter, and even the mystery of why so much ordinary matter has survived to today, physicists also joke that we live in the dark ages.

But these mysteries are precisely what make this an exciting time for anyone investigating the cosmos. Scientists have made a great deal of progress in understanding the dark sector, yet big questions remain for which we are poised to make progress. For a researcher like me, this is the optimal situation.

Perhaps one can say that physicists studying "the dark" are participating in a Copernican revolution in a more abstract form. Not only is the Earth not physically the center of the Universe, but our physical makeup is not central to its energy budget—or even to most of its matter. And, just as the first object in the cosmos that people studied was the Earth—the object with which they were most familiar—physicists focused first on the matter of which we are made, which is the most readily accessible, obvious, and essential to our lives. Exploring the geographically varying and challenging territory of the Earth wasn't always easy. But as demanding as the Earth was to fully understand, it was more accessible and easier to study than its more distant counterparts—the far-out regions of the Solar System and outer space beyond.

Similarly, discerning the most basic elements of even ordinary matter was challenging, but its study has been far more straightforward than investigating the "transparent" dark matter that is invisible—but present all around us.

However, the situation is now turning. The study of dark matter is very promising today in that it should be explained by conventional particle physics principles and furthermore should be amenable to a wide variety of currently active experimental probes. Despite the weakness of its interactions, scientists have a real chance in the coming decade of identifying and deducing the nature of dark matter. And, because dark matter clusters into galaxies and other structures, upcoming observations of the galaxy and the Universe will allow physicists and astronomers to measure it in newer ways. Furthermore, as we will see, dark matter might even account for some peculiarities of our Solar System related to meteoroid impacts and the course of life's development on Earth. Dark matter is not separated in space (and it's real) so the Starship Enterprise won't transport us there. But with the ideas and technology currently in the works, dark matter is poised to be the final frontier—or at least the next exciting one.

2

THE DISCOVERY OF DARK MATTER

When walking down the sidewalks of Manhattan or driving along the streets of Hollywood, you sometimes sense that a famous person is near. Even if you don't see George Clooney directly, the disruptive traffic generated by the waiting crowd armed with cell phones and cameras suffices to alert you to a celebrity's proximity. Though you detect the presence only indirectly, through George's substantial influence on everyone else around, you can nonetheless be confident that someone special is near.

Sometimes when you walk through a forest, a flock of birds will suddenly go wild overhead or a buck will run across your path. You might never directly encounter the hiker or hunter that set these animals in motion. Even so, the motion of the animals introduces the sportsmen and helps to tell their story.

We don't see dark matter, but—like the celebrity or the hunter—it influences its surroundings. Astronomers have used these indirect influences to infer the presence of dark matter. Today's measurements tell us about dark matter's energy contribution with ever-increasing precision. Though gravity is a weak force, sufficiently large quantities of dark matter have a measurable influence—and there is indeed a lot of it in the Universe. We don't yet know dark matter's true nature, but the measurements I'll now describe demonstrate that dark matter is a real and essential component of our world. Dark matter, though so far invisible to our eyes or direct observations, doesn't completely hide.

A BRIEF HISTORY OF DETECTING DARK MATTER

Fritz Zwicky was an independent thinker who had some impressive insights as well as some nonsensical ones. He was keenly aware of his oddball status, and even intended to write an autobiography titled *Operation Lone Wolf*. His reputation might partly explain why, even though in 1933 he made one of the most spectacular discoveries of the twentieth century, it wasn't taken seriously for another forty years.

Yet Zwicky's 1933 deduction was indeed remarkable. He observed the velocities of galaxies in the Coma Cluster (a *cluster* is a large collection of gravitationally bound galaxies). The gravitational attraction of the matter within a cluster competes with the kinetic energy of the stars it contains to create a stable system. With too low a mass, the gravitational attraction of the cluster won't prevent the stars' kinetic energy from driving them away. Based on his measurements of the velocity of the stars, Zwicky calculated that the amount of mass required for the cluster to have sufficient gravitational pull was 400 times greater than the contribution of the measured luminous mass—the matter that emits light. To account for all that extra matter, Zwicky proposed the existence of what he named *dunkle Materie,* which is German for dark matter and sounds either more ominous or sillier depending on how you pronounce it.

The brilliant and prolific Dutch astronomer Jan Oort came to a similar conclusion about dark matter a year earlier than Zwicky. Oort recognized that the velocities of stars in our local galactic neighborhood were too high for their motion to be attributed solely to the gravitational influence of light-emitting matter. Oort too deduced that something was missing. He didn't conjecture a new form of matter, however, but merely nonluminous ordinary stuff—a proposal that has since been rejected for several reasons that I'll discuss below.

But Oort might not have been the first to make this discovery either. At a cosmology conference that I recently attended in Stockholm, my Swedish colleague Lars Bergstrom told me about the relatively unknown observations of the Swedish astronomer, Knut Lundmark, who had observed matter missing from galaxies two years earlier even than Oort. Although Lundmark, like Oort, hadn't made the more daring suggestion of an entirely new form of matter, his measurements for the ratio of dark matter to visible matter most closely approximated the true value, which we now know to be about five.

Yet despite these early observations, dark matter for a long time was essentially ignored. The idea was resurrected only in the 1970s when astronomers observed the motion of satellite galaxies—small galaxies in the vicinity of larger ones—that could be explained only by the presence of additional, unseen matter. These and other observations began to turn dark matter into a topic of serious inquiry.

But its status was truly solidified by the work of Vera Rubin, an astronomer at the Carnegie Institution of Washington, who worked with the astronomer Kent Ford. After receiving her graduate degree from Georgetown University, Rubin decided to measure the angular motion of stars in galaxies—starting with Andromeda—in part to avoid trespassing on other scientists' overly protected ground. She changed the direction of her research after her thesis—which measured galaxy velocities and confirmed the existence of galaxy clusters—was initially rejected by most of the scientific community, in part for the ungallant reason that it trod on others' scientific domain. For her postgraduate work, Rubin decided to enter a less crowded research field, so she decided to study the orbital speeds of stars.

Rubin's decision led to what is perhaps the most exciting discovery of her time. In the 1970s, Rubin and her collaborator Kent Ford found that the rotational velocities of stars were pretty much the same at any distance from the galactic center. That is, stars rotated with constant velocity, even well beyond the region containing luminous matter. The only possible explanation was some as-yet unaccounted-for matter that helps rein in the farther-out stars, which moved far more quickly than expected. Without this additional contribution, the stars with the velocities that Rubin and Ford measured would fly off out of the galaxy. These researchers' remarkable deduction was that ordinary matter accounted for only about a sixth of the mass that was required to keep them in orbit. Rubin and Ford's observations yielded the strongest evidence at the time for dark matter, and galaxy rotation curves have continued to be an important clue.

Since the 1970s, evidence for dark matter, and the proportion of the Universe's net energy density it carries, has become even stronger and much better determined. The dynamical effects that allow us to learn about dark matter include the rotation of stars in galaxies that I just described. However, those measurements applied only to spiral galaxies—galaxies, like our Milky Way, that have visible matter in a disk with spiral arms extending outward. Another important category is the elliptical galaxy, in which the luminous matter has a more bulbous shape. In elliptical galaxies, as with Zwicky's measurement of galaxy clusters, one can measure *velocity dispersions*—how much velocities vary among the stars in the galaxies. Since these velocities are determined by the mass inside a galaxy, they serve as proxies for a measurement of a galaxy's mass. Measurements in elliptical galaxies further demonstrated that luminous matter is insufficient to account for the measured dynamics of their stars. On top of that, measurements of the dynamics of interstellar gas—gas not contained in stars—also called for dark matter. Because these particular measurements were made 10 times farther from the galaxies' centers than the extent of visible matter, they showed that not only did dark matter exist, its range extended far beyond the visible part of a galaxy. X-ray measurements of the gas's temperature and density confirmed this result.

GRAVITATIONAL LENSING

The mass of galaxy clusters can also be measured by the *gravitational lensing* of light. (See Figure 1.) Once again, remember that no one sees dark matter itself. But dark matter can nonetheless influence

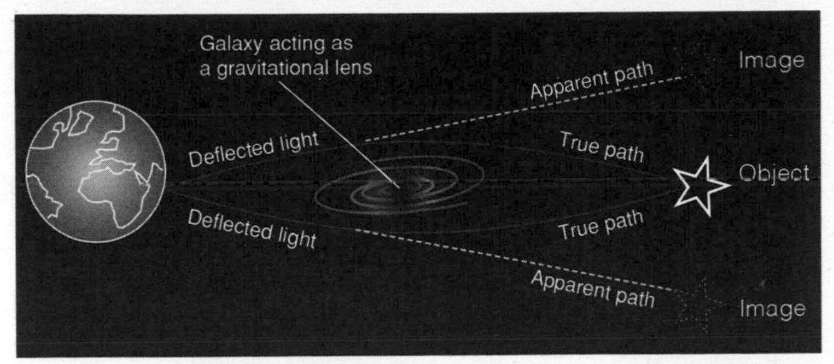

[FIGURE 1] A bright object such as a star or galaxy emits light that bends around a massive object such as a galaxy cluster. An observer on Earth projects the light as multiple images of the original light-emitting source.

surrounding matter, and even light, through its gravitational sway. For example, according to Zwicky's Coma Cluster observations, dark matter affected the motion of galaxies in ways that he could observe. Dark matter, though invisible in isolation, could be measured through its influence on visible objects.

The idea behind the gravitational lensing proposal, which the versatile Fritz Zwicky was the first to suggest, was that the gravitational influence of dark matter would also change the path of light emitted by a luminous object elsewhere. The gravitational influence of an intervening massive object such as a galaxy cluster bends the paths of the light rays that are emitted by the luminous object. When the cluster is sufficiently massive, the distortion in the paths is observable.

The direction of reorientation depends on the initial direction of the light: light that passed over the top of the cluster bends downward whereas light on the right bends to the left. Retracing the rays as if they had arrived in straight lines, observations would produce multiple images of whatever object it was that had generated the light in the first place. Zwicky realized that dark matter in galaxy clusters could thereby be detected by the observed change in light rays and the apparent multiple images, which would depend on the total mass that the intervening cluster contained. *Strong gravitational lensing* yields the multiple images of the light-emitting object. *Weak gravitational lensing*, where shapes are distorted but not duplicated, can be used at the edge of the cluster where the influence is not as pronounced.

As with the velocities of galaxies in a cluster that led to Zwicky's first radical conclusion, lensed light would yield visible consequences of the total mass of the cluster, despite the invisibility of dark matter itself. This dramatic observational consequence has been observed—but only many years after it was first suggested.

However, lensing measurements are now some of the most important observations for the study of dark matter. Gravitational lensing is exciting because it is (in a sense) a way to view the dark matter directly. The dark matter between an emitting object and the viewer bends light. This occurs independently of any dynamical assumptions like those used with star or galaxy velocity measurements. Lensing directly measures the mass between the light emitter and us. Something behind a galaxy cluster (or other dark matter containing object) emits light along our line of sight and the galaxy cluster bends that light. Lensing has also been used to measure dark matter in galaxies, with quasar light emitted behind the galaxy appearing in multiple images due to the distortion from the gravitational effects of galactic matter—including the dark, nonluminous matter.

THE BULLET CLUSTER

Measurements of gravitational lensing also play a role in what is perhaps the most compelling evidence for dark matter, which comes from clusters of galaxies that have merged—as happened with the now-famous (among physicists at least) *Bullet Cluster*. (See Figure 2.) The Bullet Cluster was actually formed from the merging of at least two galaxy clusters. The progenitor clusters contained dark matter and ordinary matter—namely X-ray emitting gas. Gas experiences electromagnetic interactions and this is enough to effectively keep the gas of the two clusters from continuing to move past each other, with the result that the gas that was initially moving along with the clusters gets gridlocked in the middle. The dark matter, on the other hand, interacts very little—both with the gas and, as the Bullet Cluster demonstrates, with itself. It could therefore flow through unimpeded, resulting in the bulbous Mickey Mouse ear shapes in the outside regions of the merged cluster. The gas acts like the traffic jam of crowded cars after two lanes have merged from different directions, whereas the dark matter is like svelte, freely moving mopeds that can race right on through.

[FIGURE 2] Clusters merge to create the Bullet Cluster, in which gas gets trapped in the central merging region and dark matter passes through, creating bulbous outer regions that contain the dark matter.

Astronomers used gravitational lensing measurements to determine that the dark matter can be found in the outer regions, while they used X-ray measurements to establish that the gas remains in the center. This is perhaps the strongest evidence available that dark matter is just that. Although people continue to speculate about modifications of gravity, it's difficult to explain the distinctive structure of the Bullet Cluster (and other similar observations) without some non-interacting matter to account for the funny shape. The Bullet

Cluster, and other similar clusters, demonstrates in the most direct manner that dark matter exists. It's the stuff that passed through unimpeded when clusters merged.

DARK MATTER AND THE COSMIC MICROWAVE BACKGROUND

The above observations established the existence of dark matter. But they still leave us with the question of what is the total energy density of the dark matter in the Universe. Even if we know how much dark matter is contained in galaxies and in galaxy clusters, we don't necessarily know the total amount. It's true that most dark matter should be tied up in galaxy clusters since a distinctive property of any type of matter is that it clumps. So dark matter should be found in gravitationally bound structures and not diffusely spread throughout the Universe, so that amount of dark matter contained in clusters should very closely approximate the total amount. Nonetheless, it would be nice to measure the energy density carried by dark matter without making this assumption.

Indeed, an even more robust way of measuring the total amount of dark matter exists. The amount of dark matter influenced the cosmic microwave background—the relic radiation from the Universe's earliest moments. The properties of this radiation, which have been measured with great precision, now play a critical role in establishing the correct cosmological theory. The best gauge of the amount of dark matter comes from studying this radiation, which is the cleanest probe into the Universe's early stages that exists.

I'll warn you in advance that the calculations are subtle—even for physicists. However, some of the essential concepts that enter the analysis are far more straightforward. A crucial piece of information is that in the beginning, atoms—which are electrically neutral bound states of positively charged nuclei and negatively charged electrons—didn't exist. Electrons and nuclei could stably combine into atoms only after the temperature dropped below the atomic binding energy. Above that temperature, radiation would separate protons and electrons and would thereby blow atoms apart. Because of all the charged particles early on, radiation permeating the Universe initially didn't travel freely. Instead, it scattered off the many charged particles that the early Universe contained.

However, as the Universe cooled, charged particles combined to form neutral atoms at a particular temperature known as the recombination temperature. This absence of unbound charged particles gave photons free license to travel undeterred. As a result, from this time forward, charged particles no longer traveled independently and instead became bound into atoms. With no charged particles off which to scatter, the photons emitted after recombination could head straight to our telescopes. So when we look at the cosmic microwave background radiation, we are looking back to that relatively early time.

From a measurement standpoint, this is fantastic. This was early enough in the Universe's lifetime—about 380,000 years after the Big Bang—that structure hadn't yet formed. The Universe took more or less the simple form that our initial cosmological picture suggests. Mostly it was homogeneous and isotropic. That is, the temperature was almost the same no matter which part of the sky you examined or which direction you chose. But minor fluctuations in temperature at the level of one part in ten thousand slightly compromised this homogeneity. The measurements of these fluctuations carry an enormous amount of information about the Universe's contents and subsequent evolution. The results allow us to deduce the Universe's expansion history and other properties that tell us about the amount of radiation, matter, and energy present, then and today, yielding detailed insights into the properties and contents of the Universe.

To have some understanding of why this ancient radiation is so rich with information, the second thing to appreciate about the early Universe is that at recombination time, when neutral atoms could finally form, the matter and radiation in the Universe began to oscillate. In what is known as *acoustic oscillations,* the gravitational pull of matter pulled stuff in while the pressure of radiation drove stuff out. These forces competed with each other, leading collapsing matter to contract and expand, creating oscillations. The amount of dark matter determined the strength of the gravitational potential that pulled stuff in, resisting the outward force of the radiation. This influence helped shape the oscillations, allowing astronomers to measure the total dark matter energy density that was present at the time. In an even more subtle effect, dark matter also influenced how much time elapsed between when matter began to collapse (which happens when the energy density in matter exceeds that in radiation) and the time of recombination—when stuff began to oscillate.

THE COSMIC PIE

This is a lot of information. But even without knowing the details, we can understand that these measurements are extraordinarily precise, and that they let us pin down very accurate values for many cosmological parameters—including the net amount of energy density carried by dark matter. The measurements not only confirm the existence of dark energy and dark matter—they constrain the fraction of the Universe's energy they contain too. The percentage of energy in dark matter is about 26 percent, in ordinary matter about 5 percent, and in dark energy about 69 percent. (See Figure 3.) Most of the energy of ordinary matter is carried by atoms, which is why the cosmic pie chart uses "atoms" and "ordinary matter" interchangeably. In other words, dark matter carries five times the energy of ordinary matter, meaning it carries 85 percent of the energy of matter in the Universe. It was reassuring that the result for the dark matter contribution which we derive from the measured background radiation turned out to agree with the previous measurements from galaxy clusters—thus reinforcing the result obtained from the microwave background measurements.

The cosmic microwave background measurements also confirmed the existence of dark energy. Because dark matter and ordinary matter affect in different ways the perturbations in the radiation background—the radiation that survives today from the time of the Big Bang—data about this background confirmed the existence of dark matter and furthermore measured how much of it is present. However, dark energy—the mysterious form of energy that the previ-

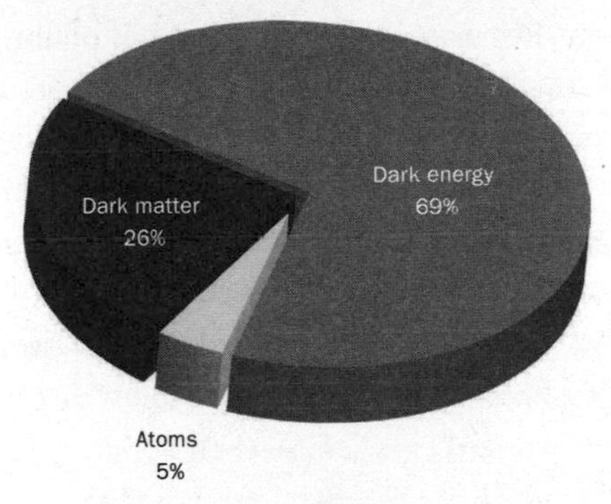

[FIGURE 3] Pie chart illustrating the relative amounts of energy stored in ordinary matter (atoms), dark matter, and dark energy. Notice that dark matter is 26 percent of the total energy density, but makes up about 85 percent of the energy of matter because matter includes only the atom and dark matter contributions, but not that of dark energy.

ous chapter discussed, which is present in the Universe but not carried by any form of matter—influences these fluctuations too.

But the true discovery of dark energy came from supernova measurements, made by two separate teams of physicists—one led by Saul Perlmutter and the other by Adam Riess and Brian Schmidt. This discovery is a minor detour since dark matter will be our true concern. But dark energy too is interesting and important enough to be worth a brief detour here.

TYPE IA SUPERNOVAE AND THE DISCOVERY OF DARK ENERGY

Type Ia supernovae played an especially important role in the discovery of dark energy. Type Ia supernovae result from the nuclear explosions of *white dwarfs,* which are the seemingly innocuous end states of some stars' evolution after they have burned up all the hydrogen and helium at their core via thermonuclear fusion. Above a certain mass, white dwarfs become unstable and explode. Like an oil-rich country that has exported away all its resources and finds itself saddled with a dense, dissatisfied population on the brink of revolution, white dwarfs absorb only so much material before their mass sets them on the verge of exploding too.* Because the white dwarfs that explode to produce them all have the same mass, type Ia supernovae all shine with about the same brightness, making them what astrophysicists refer to as *standard candles*.†

Type Ia supernovae are a particularly useful probe of the Universe's expansion rate because of this uniformity and also because they are bright and relatively easy to see, even when far away. On top of that, they are standard candles, so their apparent brightness varies only because of their different distances from us.

So if astronomers measure both the speed with which a galaxy is receding and the galaxy's luminosity, they can determine the expansion rate of the Universe in which the galaxy is carried, as well as how far away the galaxy is. Armed with this information, they can determine the Universe's expansion as a function of time.

Two teams of supernova researchers used this insight to discover dark energy in 1998, when they measured the redshifts of the galaxies in which the Type Ia supernovae reside. The redshift is the shift in the frequency of the light that a receding object emits, which, like the lowering of the pitch of an ambulance's siren when it's racing away, tells how quickly something moves from a source of light or

---

* Admittedly, this analogy goes only so far. Unlike disgruntled exiles, heavy elements—once distributed around the Universe—won't spur further instabilities. Better still, they contribute to the formation of stellar systems and even to life.

† In fact, this picture—though widely accepted—is currently disputed by experts. On the one hand, predictions for the spectra and light curves of exploding white dwarfs match the observations very well. On the other hand, no one has seen the expected companion binary star to the white dwarfs. Astronomers therefore suggest that it might instead be the merging of two white dwarfs that gives rise to the explosion. Some data support this conclusion—mostly having to do with measuring the difference in time between the binary's formation and its explosion—but the detailed predictions for the single white dwarf explosion scenario are yet to be confirmed, so the matter remains unresolved.

sound. Knowing both the redshifts and the brightness of the supernovae they studied, researchers could measure the Universe's expansion rate.

Much to their surprise, they found that supernovae were dimmer than anticipated, indicating that the supernovae were farther away than they had expected according to then-conventional assumptions about the Universe's expansion rate. Their observations led to the remarkable conclusion that some unanticipated energy source was accelerating the rate of the Universe's expansion. Dark energy fits the bill, since its gravitational influence makes the Universe expand at an increasingly rapid rate over time. Along with measurements in the cosmic microwave background, the supernovae measurements established the existence of dark energy.

DARK MATTER CODA

Today the agreement among all measurements is so good that cosmologists talk of a ΛCDM paradigm, where Λ refers to "Lambda" and CDM is an acronym for cold dark matter. Lambda is the name that is sometimes used for the dark energy that we know now to be present. With dark energy, dark matter, and ordinary matter distributed as in the cosmic pie, all measurements to date agree with predictions.

The precisely measured cosmic microwave background radiation with its minuscule but rich density perturbations establishes many cosmological parameters, including the energy densities of ordinary matter, dark matter, and dark energy—as well as the Universe's age and shape. The excellent agreement of the more recent data from the WMAP and Planck satellites that we will consider in Chapter 5, and with the data from other observations such as that obtained by studying Type 1a supernovae, is an important verification of the cosmological model.

But there is one final, very important piece of evidence for the existence of dark matter that I have yet to address. This one—probably the most important to us—is the existence of structure such as galaxies. Without dark matter, such structure wouldn't have had time to form.

To appreciate dark matter's pivotal role in this important process requires some understanding of the earlier stages of the Universe's history. So before getting to the formation of structure, let's first turn to cosmology, the study of how the Universe changed over time.

3

THE BIG QUESTIONS

A couple of times when I've told people that I work on cosmology, they have mistaken me for a cosmetologist, which I find very funny given how poorly suited I would be for this vocation. But the mistake did motivate me to look up these words, which—if you are not listening closely—can sound strikingly similar. The Online Etymology Dictionary, which explained that both words originated in the Latinized version of the Greek *kosmos,* taught me that the mistake is almost justifiable. Pythagoras of Samos in the sixth century B.C. might have been the first to use *kosmos* to apply to the Universe. But from around A.D. 1200, the meaning of *cosmos* was "order, good order, or orderly arrangement." The word didn't gain traction, however, until the mid-nineteenth century, when the German scientist and explorer Alexander von Humboldt gave a series of lectures, which he wrote up in a treatise titled *Kosmos*. This treatise influenced many readers, including the writers Emerson, Thoreau, Poe, and Whitman. You might joke that Carl Sagan did the original rebooting of the popular *Kosmos* series.

The word *cosmetic,* on the other hand, traces back to the 1640s—derived from the French *cosmétique,* which in turn descended from the Greek *kosmetikos,* meaning "skilled in adornment or arrangement." The online dictionary, presenting a dual meaning that I suspect only Los Angeles residents can fully understand, explains "thus kosmos had an important secondary sense of 'ornaments, a woman's dress, decoration' as well as 'the Universe, the world.'" In any case, the similarity—and embarrassing confusion—that I encountered was not entirely a coincidence. Both "cosmology" and "cosmetology" derive from *kosmos*. Like a face, the Universe has both beauty and an underlying order.

Cosmology—the science of the evolution of the Universe—has now truly come into its own. It has recently entered an era in which revolutionary advances—both experimental and theoretical—have yielded a more extensive and detailed understanding than most people would have been thought possible even thirty years ago. Improved technology combined with theories rooted in general relativity and particle physics have provided a detailed picture of the Universe's earlier stages, and of how it evolved into the Universe we currently see. The next chapter explains how far and deep these twentieth-century advances have taken us in our understanding of the Universe's history. But before exploring these remarkable achievements, I want to first briefly wax philosophical in order to make clear what science does and does not tell us about the answers to some of humanity's oldest and most fundamental questions.

QUESTIONS WITHOUT ANSWERS

Cosmology is about big inquiries—nothing less than how the Universe began and subsequently developed into its current state. Prior to the scientific revolution, people tried to answer such questions with the only methods at their disposal, namely philosophy and limited observations. Some ideas they had turned out to be correct but—not surprisingly—many others were wrong.

Today too, despite our many advances, people can't help turning to philosophy when thinking about the Universe and the questions we haven't yet answered—in effect forcing us to confront the distinction between philosophy and science. Science concerns those ideas that at least in principle we can verify or rule out through experiments and observations. Philosophy, to a scientist at least, concerns questions we expect we will never reliably answer. Technology sometimes lags behind, but we'd like to believe that at least in principle scientific proposals will be verified or ruled out.

This leaves scientists with a dilemma. The Universe almost certainly extends beyond the domain we can observe. If indeed the speed of light is finite and if our Universe has been around for only a fixed amount of time, we can access only a finite region of space—no matter how much technology might advance. We can see only those regions that can be reached via a light ray—or something else that travels at light speed—during the lifetime of the Universe. Only from those regions can a signal possibly reach us within the time that the Universe has been around. Anything farther away—beyond what physicists call the *cosmic horizon*—is inaccessible to any observation we could make at this time.

This means that science in its truest form doesn't apply beyond this domain. No one can experimentally validate or rule out conjectures that apply beyond the horizon. According to our definition of science, for those faraway regions, philosophy reigns supreme. This doesn't mean that curious scientists never ponder the big questions about the physical principles or processes that apply there. Indeed, many do. I don't want to dismiss such inquiries—they can often be deep and fascinating. But given the limitations, you can't trust scientists' answers about this domain—at least not more than anyone else's. However, since I'm so often asked, I'll use this chapter to offer my take on a few of the big questions people often want addressed.

One question I frequently hear is why there is something rather than nothing. Though none of us knows the true reason, I'll give my two responses. The first, which is undeniable, is that you wouldn't be here to ask this question and I wouldn't be here to answer it if there were nothing. But my other answer is that I just think something is more likely. After all, nothing is very special. If you have a number line, "zero" is just one infinitesimal point among the infinity of possible numbers you can choose. "Nothing" is so special that without an underlying reason, you wouldn't expect it to characterize the state of the Universe. But even an underlying reason is something. You at least need physical laws to explain a very nonrandom occurrence. A cause implies there must be something. Though it sounds like a joke, I really believe this. You might not always find what you are looking for, but you don't randomly find nothing.

But there is also a scientific, rather than a philosophical question, that comes up when physicists consider the matter we are made of—the stuff we are supposed to understand. Why in our Universe do we have as much of the matter we are made of—protons, neutrons, electrons—as we do? Though we understand a great deal about ordinary matter, we don't fully understand why so much of it is still here. The amount of energy in ordinary matter is an unsolved problem. We don't yet know why it has survived as abundantly as it has to today.

The problem boils down to the question of why there wasn't always an equal amount of matter and *antimatter*. Antimatter is the stuff with the same mass but opposite charges to ordinary matter. Physical theories tell us that for every matter particle, an antimatter particle must exist. For example, knowing that an electron has charge –1 tells us there must also be an antiparticle—it's called a positron—with the same mass but opposite charge, +1. To avoid any confusion, let me state explicitly that antimatter is not dark matter. Antimatter carries the same types of charges as ordinary matter and therefore interacts with light. The only difference is that the antimatter charges are the opposite to those of the associated matter.

Because antimatter carries opposite charges to our usual matter, the net charge of matter and antimatter is zero. Since matter and antimatter together carry no charge, charge conservation and Einstein's famous formula $E = mc^2$ tell us that matter can meet up with antimatter to disappear into pure energy—which also has no charge.

We would have expected that as the Universe cooled, essentially all known matter would have annihilated with antimatter, meaning matter and antimatter would have combined to turn into pure energy, and thereby disappeared. But as we are here to discuss the question, clearly this was not the case. We are left with matter—that five percent of the Universe's energy that you see in Figure 3—so the amount of matter in the Universe must be greater than the amount of antimatter. A critical feature for our Universe—and ourselves—is that in contrast to standard thermal expectations, ordinary matter sticks around and survives in sufficient quantities to create animals and cities and stars. This is possible only because matter dominates over antimatter—there is a matter-antimatter asymmetry. If the amounts had always been equal, matter and antimatter would have found each other, annihilated, and disappeared.

In order to have matter stick around until today, an asymmetry between matter and antimatter had to have been established at some time in the early Universe. Physicists have suggested many workable scenarios for what could have created this imbalance, but we don't yet know which if any of these ideas are correct. The origin of the asymmetry remains one of the important unsolved problems in cosmology. This means that not only do we not understand the dark components, but we don't even fully understand ordinary matter—the small piece of the cosmic pie that represents known matter. Something special had to have occurred early in the evolution of the Universe to explain why this piece of pie remains.

A second currently unanswerable question is what exactly happened during the Big Bang. Scientists and the popular press frequently refer to the Big Bang explosion that happened back when the Universe was less than 10^{-43} seconds old and the Universe was 10^{-33} cm big, and even "illustrate" the explosion with gorgeous multicolored images. But the term "Big Bang" is misleading, as I will further discuss in the chapter that follows. The astronomer Fred Hoyle, who preferred a static Universe, invented the term in 1949 as a pejorative for use on his BBC radio show to refer to this theory that he didn't believe.

Regardless of your attitude toward Big Bang cosmology, which very successfully describes the Universe's evolution only a fraction of a second after the Universe we know began, no one knows what happened at that earliest moment. A reliable characterization of the Big Bang—and possibly what happened before—requires a theory of quantum gravity. On the tiny distance scales that are relevant to this earliest time, both quantum mechanics and gravity are important, and no one has yet found a solvable theory that applies to this infinitesimal distance regime. We will gain insight into the very beginning of the Universe only when we know more about physical processes on this tiny distance scale. And even then observations to validate the conclusions will very likely be impossible.

An even more impossible to answer question I often hear is, "What came before the Big Bang?" Answering this presumably requires even more knowledge than understanding the Big Bang itself. We don't know what happened at the time of the Big Bang and neither I nor anyone else knows what came before. But before you become too disappointed by this omission, let me reassure you that you would probably find any answer to this question unsatisfactory. Either the Universe was around an infinite length of time or it started at some particular time. Both answers can seem disturbing, but those are the options.

Taking this a step further, if the Universe existed forever and the Big Bang was part of it, either our Universe was all there was or other universes also emerged from their own Big Bangs. The *multiverse* is the name associated with a cosmos in which, in addition to our own universe, there are many others. In this scenario, there would be many different expanding regions—each constituting its own universe.

This reasoning leaves us with three choices. Either our universe started with the Big Bang, the Universe has been around forever but eventually went over to the expansion that the Big Bang theory predicts, or we are one of many universes that grew out of a universe/multiverse that has always existed. This covers all the possibilities. The last one seems most likely to me in that it doesn't assume that our world or even our particular universe is special, which is reasoning that has been invoked since the time of Copernicus. This choice also implies that just as the spatial extent of the universe—at least to my way of thinking—is more likely to be infinite than finite in size, the evolving universe is unlikely to have a beginning or end in time—even though our particular universe might. The existence of multiple emerging and eventually disappearing universes is probably the least unsatisfying of three not completely graspable possibilities.

This brings me to the last philosophical inquiry—prompted by the one above—which is whether such a multiverse exists. Existing physical theories suggest that multiverses are rather likely, especially given the many possible solutions in quantum gravity theories as currently formulated. Whether or not those calculations stand up to scrutiny, I would wager that other inaccessible universes should be present. Why shouldn't they? Given that we know the limitations of physical laws and current technology, it is both figuratively and literally shortsighted to decide they aren't there. Nothing about our world is inconsistent with the existence of a multiverse.

But that doesn't mean we will ever know. If nothing travels faster than the speed of light, any region that is too far away—beyond the cosmic horizon—is off-limits to observations. Yet these other regions could in principle contain other universes that are completely separate from our own. Some signals of other universes could potentially be found in the cases where over time separate universes come into contact. But this is highly unlikely, and in general other universes are inaccessible.

For my faithful, returning readers, I'll now make an aside to be clear that in discussing multiverses, I am not referring to the multidimensional scenarios that I described in *Warped Passages*. Universes might also exist that are closer than the horizon but that are separated from us across another dimension of space—a dimension beyond the three that we observe: left-right, up-down, and forward-backward. Although no one has yet seen such a dimension, it might exist and in principle a universe separated from us along this dimension might as well. This type of universe is known as a *braneworld*. As those who read my first book know, the braneworlds that interest me most could potentially have observable consequences because they are not necessarily so far away. However, braneworlds are generally not what people mean when discussing the more general multiverse scenario involving many separate universes that won't interact even via gravity. Multiverses are so far away that even something traveling at the speed of light from one of these other universes wouldn't have time to reach us in the lifetime of our Universe.

Nonetheless, there is a lot of interest in the multiverse idea in the popular imagination. I was recently talking to a friend who was very excited about the idea of a multiverse and didn't understand why I didn't necessarily find it as interesting as he did. For me the first reason is the one stated above: in all likelihood, we will never know with certainty whether or not we live in a multiverse. Even if other universes exist, they are likely to remain undetectable. My friend found this only mildly disappointing and his interest persisted. I suspect that he—along with many others—likes the idea because he thinks a copy of himself is living in one of these distant realms. Just for the record, I don't hold this view. If other universes exist, they are most likely nothing like our own. They probably don't even contain the same forms of matter or forces as we do. If there were life there, we most likely wouldn't even recognize it and probably couldn't detect it in the first place—even if it weren't so far away. The infinite number of confluences that create any single human being would be even less likely. After I explained how—even with many other universes—there can be an even larger universe of possibilities, my friend began to see my point.

In fact, even if the multiverse scenario holds, most of the other universes will be untenable and will either collapse or explode, in which case they will dilute to nothing almost instantaneously. Only a few, like our own, might last long enough to develop structure and perhaps even life. Despite Copernicus's insightful perspective, our particular Universe does seem to have a number of peculiar properties—ones that permit galaxies, the Solar System, and life. Some people try to explain the special properties of our Universe by assuming the existence of multiple universes, at least one of which has the special properties we require for our existence. Many who think this way attempt *anthropic reasoning*, which tries to justify particular properties of our Universe on the grounds that they are essential to life—or at least galaxies that can support life. The problem here is that we don't know which properties are anthropically determined and which are based on fundamental physical laws, or which properties are essential to life and which are simply essential to the life that we see. Anthropic reasoning might be correct in some cases, but we have the usual problem that we don't know how to test the ideas. In all likelihood, we will only rule out such ideas if a better, more predictive idea takes its place.

Ideas such as those discussed above are speculations. They are intriguing, but we won't have answers—at least not anytime soon. In my research, I prefer to think about the "multiverse" of communities of matter that's right here and which we can hope to understand. I'm using the term metaphorically, but it's not so far from the truth. A universe of dark matter is right under our noses. Yet we don't generally interact with it and we don't yet know what it is. But theoretical and experimental physicists are currently advancing our knowledge about what this "dark universe" might be. Someday soon we might know the answer, and such a discovery would be worth the wait.

4

ALMOST THE VERY BEGINNING: A VERY GOOD PLACE TO START

A rather funny and outspoken Russian theoretical physicist startled everyone over coffee recently when he was describing the colloquium he was planning for the following week. A physics colloquium is a general talk that is geared toward students, postdoctoral fellows, and professors—all of whom have a background in physics, though not necessarily with a focus in the speaker's more narrowly defined field. This particular physicist's description of his proposed colloquium was "I will talk about cosmology." When it was pointed out this might be a bit broad—after all, cosmology is an entire discipline—he argued that there are only a few ideas and quantities worth measuring in cosmology and he could cover all of them—along with his own contributions—in an hour-long talk.

I'll let you be the judge of whether this extreme view of cosmology is actually true—for the record I'm doubtful. Many issues remain to be explored and understood. But indeed, part of the beauty of the Universe's early evolution is that in many respects it is surprisingly simple. By looking at the single sky that astronomers and physicists observe and study today, we can extrapolate facts about the composition and activities of the Universe billions of years ago. In this chapter, we'll explore the stunning progress in our understanding of the Universe's history that the past century's beautiful theories and measurements have brought us.

THE BIG BANG THEORY

We don't have the tools to reliably characterize the very beginning. But not knowing how the Universe started doesn't mean that we don't know quite a lot. Unlike its very beginning, which no known theory can describe, the Universe's evolution only a tiny fraction of a second after its beginning hewed to established laws of physics. By applying the equations of relativity and using simplifying assumptions about the Universe's contents, physicists can determine a great deal about the Universe's behavior only a minuscule interval after it began—perhaps 10^{-36} seconds or so later—at which time the Big Bang theory, which describes the expansion of the Universe, applies. The Universe at this early time was filled with matter and radiation that was uniform and isotropic—the same at all places and in all directions—so only a few quantities suffice to describe its early physical properties. This characterization makes the early evolution of the Universe simple, predictable, and understandable.

The lynchpin of the Big Bang theory is the Universe's expansion. In the 1920s and 1930s, the Russian meteorologist Alexander Friedmann, the Belgian priest and physicist Georges Lemaître, the American mathematician and physicist Howard Percy Robertson, and the British mathematician Arthur Geoffrey Walker—the last two working together—solved Einstein's equations of general relativity and deduced that the Universe must grow (or contract) with the passage of time. They moreover calculated how the expansion rate of space would respond to the gravitational influence of matter and radiation, both of whose energy densities change too as the Universe evolves.

The expansion of the Universe is perhaps an odd concept, given that the Universe has very likely been infinite all along. But it is space itself that is expanding, meaning that the distances between objects like galaxies increases with time. I am frequently asked, "If the Universe is expanding, what is it expanding into?" The answer is that it is not expanding into anything. Space itself grows. If you imagine the universe as the surface of a balloon, the balloon itself stretches. (See Figure 4.) If you had marked two points on the balloon's surface, those two points would grow farther apart, just as galaxies recede from each other in an expanding universe. Our analogy isn't perfect since the surface of the balloon is only two-dimensional and it does in fact expand into three-dimensional space. The analogy works only if you imagine that the balloon's surface is all there is—it is space itself. If this were true—even with nothing else already there to expand into—the marked points would nevertheless still grow apart.

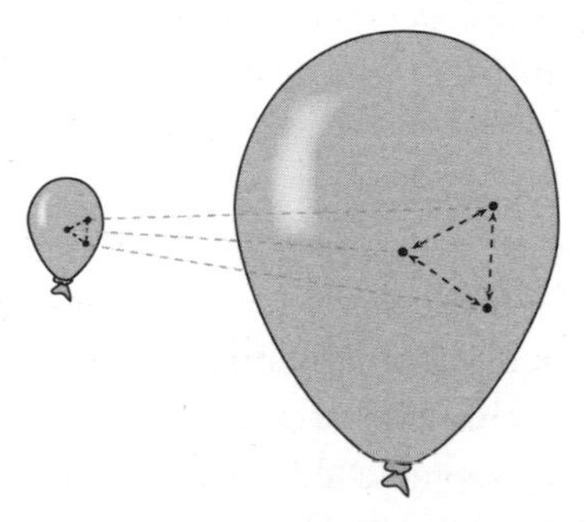

[FIGURE 4] Galaxies move away from each other as the Universe expands, much as points on a balloon move away from each other as it is blown up.

THE BALLOONIVERSE

For the analogy to be even better, only the space between the marked points would expand—not the points themselves. Even in an expanding universe, stars or planets or anything else that is sufficiently tightly bound by other forces or by stronger gravitational effects won't experience the expansion that drives galaxies apart from each other. Atoms, which consist of a nucleus and electrons that are kept in close proximity by the electromagnetic force, don't get any bigger. Neither do relatively dense strongly bound structures such as galaxies—or our own bodies for that matter, which have a density in excess of a trillion times the mean density of the Universe. The force driving the expansion acts on all these dense bound systems too, but because other force contributions are so much more powerful, our bodies and the galaxies don't grow with the Universe's expansion—or if they do, it is by such a negligible amount that we would never notice or measure the effect. Matter that is bound more strongly than the force driving the expansion remains the same size. It is only the distance between such objects that gets bigger as the growing space drives them farther apart.

Einstein famously first derived the expansion of the universe from his equations of relativity. He did so, however, before any expansion was measured, so he didn't accept or advocate his result. In an attempt to reconcile his theory's predictions with a static universe, Einstein introduced a new source of energy that looked to him like it could thwart the predicted expansion. Edwin Hubble proved this kludge misguided in 1929 when he discovered that the Universe was indeed expanding, with galaxies moving farther away from each other over time (though incredibly, as an observer who didn't trust any particular theory, he didn't accept this interpretation of his results). Einstein readily dispensed with the fudge he had made, and is known (perhaps apocryphally) to have called it his "biggest blunder."

The modification was not entirely erroneous, however, in that the type of energy that Einstein proposed does exist. More recent measurements have shown that the new type of energy he added, which we now call "dark energy"—though not of the magnitude or of a type that would curb the expansion of the Universe—is actually necessary to account for recent observations of precisely the opposite effect—the Universe's accelerated expansion. But I think Einstein really thought his blunder—if indeed he actually called it that—was in not recognizing the correctness and significance of his initial expansion prediction, which could have been viewed as a key prediction of his theory.

In fairness, before Hubble presented his results, very little was known about the Universe. Harlow Shapley had measured the size of the Milky Way to be 300,000 light-years across, but he was convinced that the Milky Way was all that the Universe contained. In the 1920s, Hubble realized that this was not true when he discovered that many nebulae—which Shapley had thought were clouds of dust that merited this uninspiring name—were in fact other galaxies, millions of light-years away. Toward the end of the decade, Hubble made his even more famous discovery—the *redshift of galaxies*—the shift in frequency of light that told scientists the Universe was expanding. The redshift of galaxies—like the lowering in pitch of a moving ambulance's siren that tells us it is moving away—demonstrated that other galaxies were receding from ours, indicating that we live in a Universe in which galaxies grow farther apart.

Today, we sometimes talk about a Hubble constant, which is the rate at which the Universe currently expands. It is a constant in the sense that today, its value everywhere in space is the same. But actually the Hubble parameter is not constant. It changes with time. Earlier in the Universe, when things were denser and gravitational effects were stronger, the Universe expanded far more rapidly than it does today.

Until rather recently, a fairly broad range of "measured" values for the Hubble parameter, which quantifies the expansion rate today, meant we could not precisely pin down the Universe's age. The lifetime of the Universe depends on the inverse of the Hubble parameter, so if that measurement is uncertain by a factor of two, so too is its age.

I remember reading in the newspaper when I was a kid that some recent measurements had caused the age of the Universe to be revised by this amount. Not knowing this represented the expansion rate measurement, I remember my astonishment at the radical revision. How could something as important as the age of the Universe have been at liberty to change? It turns out we can understand a good deal of the Universe's evolution at a qualitative level, even without knowing its precise age. But better knowledge of its age does foster a better understanding of the Universe's contents and the underlying physical processes at work.

This uncertainty is in any case now under much better control. Wendy Freedman, who was then at the Carnegie Observatories, and her collaborators measured the expansion rate and ultimately quelled the debate. In fact, because the value of the Hubble parameter is so important for cosmology, a concerted effort was made to ensure the greatest accuracy possible. Using the Hubble Space Telescope (given the name, it seems only fair), astronomers measured a value of 72 km/sec/Mpc (meaning something at a distance of a megaparsec moves away at 72 km/sec) with an accuracy of 11 percent—a far cry from Hubble's original and very inaccurate measurement of 500 km/sec/Mpc.

A megaparsec (Mpc) is a million parsecs, and a parsec, like many astronomical units, is a historical relic from the way distances were measured in early times. It is a shortened version of "parallax second" and has to do with the angle subtended by an object on the sky, which is why it has an angular unit in it. Although many astronomers still use the units, as they do with many other nonintuitive, historically motivated measures, most people prefer not to think in terms of parsecs. To convert to what is perhaps a slightly more familiar measure of distance, a parsec is about 3.3 light-years. It is a fortuitous coincidence that the arcane measure is roughly equivalent to the more readily interpreted quantity.

The Hubble Telescope's more accurate result for the Hubble parameter might have been uncertain by 10 to 15 percent, but it was not uncertain by a factor of two. More recent results relying on measurements of the cosmic microwave background radiation data do even better. The age of the Universe is now known to within a couple of hundred million years, and measurements have continued to improve. When I wrote my first book it was 13.7 billion years old but we now believe it to be a bit older—13.8 billion years from the so-called Big Bang. Note that it is not only the changing Hubble parameter, but the discovery of the dark energy that I mentioned in Chapter 1, that led to this more refined result, since the age of the Universe depends on both.

PREDICTIONS OF BIG BANG EVOLUTION

According to the Big Bang theory, the very early Universe originated 13.8 billion years ago as a hot, dense fireball consisting of many interacting particles with temperature higher than a trillion trillion degrees. All known (and presumably as-yet unknown) particles zipped around everywhere at close to the speed of light, constantly interacting, annihilating, and being created from energy in accordance with Einstein's theory. All types of matter that interacted sufficiently strongly with each other had a common temperature.

Physicists call the hot, dense gas that filled the Universe in its early stages *radiation*. For cosmological purposes, radiation is defined as anything that moves at relativistic speeds, which means at or very close to the speed of light. To count as radiation, objects have to possess so much momentum that their energy far exceeds the energy stored in their mass. The early Universe was so outrageously hot and energetic that the gas of fundamental particles that comprised it readily satisfied this criterion.

Only fundamental particles were present in this Universe, and not, for example, atoms, which are made of nuclei bound together with electrons—or protons—that are made from the more fundamental particles called quarks. Nothing could remain trapped in a bound object in the face of so much heat and energy.

As space expanded, the radiation and particles that permeated the Universe became more dilute and cooled down. They behaved like hot air trapped inside a balloon, which becomes less dense and cooler as the balloon expands. Because each energy component's gravitational influence affects the expansion differently, the study of the Universe's expansion over time allows astronomers to disentangle the separate contributions of radiation, matter, and dark energy. Matter and radiation dilute with the expansion but radiation, which redshifts to lower energy—much like a siren decreases in frequency as it moves away—dilutes even more rapidly than matter. Dark energy, on the other hand, doesn't dilute at all.

As the Universe cooled, notable events occurred when its temperature and energy density no longer sufficed to produce a particular particle. This happened at times when a particle's kinetic energy no longer exceeded mc^2, where m is the mass of that particular particle and c is the speed of light. One by one, massive particles became too heavy for the cooling Universe. By combining with antiparticles, such heavy particles annihilated, converting into energy that then heated up the remaining light particles. The heavy particles thereby decoupled and essentially disappeared.

But even though the Universe's contents changed, nothing observable happened until a few minutes into the Universe's Big Bang evolution. So we'll jump ahead to when the contents of the Universe changed substantially—and did so in a verifiable way. The Hubble expansion mentioned above was one confirmation of the Big Bang theory. Two other significant measurements—both involving the Universe's contents—solidified physicists' confidence that it was correct. We'll first consider the prediction of the relative fractions of the different types of nuclei that were formed in the very early Universe, which match pretty closely the densities that have been observed.

A few minutes after the "Big Bang," protons and neutrons stopped flying around in isolation. The temperature dropped sufficiently that these particles became bound into nuclei in which they were held together by strong nuclear forces. Also by that time, matter interactions that initially kept the number of protons and neutrons the same were no longer effective. Because neutrons could still decay into protons through the weak nuclear force, their relative number changed.

Because neutron decay takes place sufficiently slowly, a substantial fraction of neutrons survived long enough to be absorbed into nuclei along with the protons that were present. Helium, deuterium, and lithium nuclei were then formed and the cosmic relic amount of these elements, as well as of hydrogen—whose density was depleted when helium was created—was established. The residual amounts of different elements were set by the relative number of protons and neutrons as well as by how quickly the required physical processes took place relative to the speed at which the Universe expanded. So the predictions of *nucleosynthesis* (as this process is known) test the theory of nuclear physics as well as the details of the Big Bang expansion. In a significant confirmation of both the Big Bang theory and nuclear physics, observations agree with predictions spectacularly well.

Not only did these measurements verify existing theories, but they constrain new ones too. That is because the expansion rate when nuclei abundances were established is mostly accounted for by the energy carried by the types of matter we already know about. Whatever new stuff existed at the time better not have contributed too much energy back then or the expansion rate would have been too rapid. This constraint is important for me and my colleagues when we consider more speculative ideas for what can exist in the Universe. Only small amounts of novel forms of matter could have been in equilibrium and had the same temperature as known matter at the time of nucleosynthesis.

The success of these predictions also tells us that, even today, the amount of ordinary matter cannot be much greater than what has been observed. Too much normal matter and the predictions of nuclear physics wouldn't match the observed heavy element abundances in the Universe. Along with the measurements described in the previous chapter, which tell us that luminous matter does not suffice to explain observations, the successful predictions of nucleosynthesis tell us that ordinary matter cannot account for all the observed matter in the Universe—largely dispelling the hope that it was invisible just because it wasn't burning or reflective enough. If there were much more ordinary matter than is observed in luminous matter, the successful nuclear physics predictions would no longer apply unless there were some new ingredient. Unless ordinary matter can somehow hide during nucleosynthesis, we have to conclude that dark matter must exist.

But perhaps the most significant milestone in the Universe's evolution, at least in terms of detailed testing of cosmological predictions, occurred somewhat later on—about 380,000 years after the Big Bang. The Universe was originally filled with both charged and uncharged particles. But at this later time, the Universe had cooled sufficiently that positively charged nuclei combined with negatively charged electrons to form neutral atoms. From that time forward, the Universe consisted of neutral matter, which is matter that carries no electric charge.

For *photons,* the particles that communicate the force of electromagnetism, this sequestering of charged particles into atoms was a substantial change. In the absence of charged matter to deflect them, photons could traverse the Universe unhindered. This meant that radiation and light from the early Universe could reach us directly, essentially independent of any more complicated evolution in the Universe that might occur later on. The background radiation we see today is the radiation that existed 380,000 years into the Universe's evolution.

This radiation is the same radiation that was present immediately after the Universe began its Big Bang expansion, but it is now at a much lower temperature. The photons cooled, but they didn't disappear. The temperature of the radiation today is 2.73 kelvin,* which is extremely cold. The radiation's temperature is only a few degrees warmer than zero kelvin—also known as absolute zero, the coldest anything can be.

The detection of this radiation was in some sense the "smoking gun" for the Big Bang theory, perhaps the most convincing evidence that the equations were correct. The German-born astronomer Arno Penzias and the American Robert Wilson accidentally discovered this cosmic microwave background radiation in 1963, while using a telescope at Bell Labs in New Jersey. Penzias and Wilson weren't actually looking for cosmological relics. They were interested in radio antennas as a way to do astronomy. Of course, Bell Labs, which was associated with a telephone company, was interested in radio waves too.

But when Penzias and Wilson tried to calibrate their telescope, they recorded a uniform background noise (like static) that came from all directions and didn't change with the seasons. It never went away so they knew they couldn't ignore it. Because it had no preferred direction, it couldn't be coming from nearby New York City, the Sun, or the test of a nuclear weapon the previous year. After cleaning off the droppings from the pigeons nesting inside the telescope, they concluded it couldn't come from the pigeons' "white dielectric material," as Penzias politely called it, either.

---

* Temperature differences in kelvin are identical to differences measured in degrees Celsius, but the minimum possible temperature is 0, rather than −273.15, as it is in Celsius units, or −459.67, as it is in Fahrenheit ones.

Robert Wilson told me the story of how lucky they were in the timing of their discovery. They didn't know anything about the Big Bang, but the theoretical physicists Robert Dicke and Jim Peebles at nearby Princeton University did. The Princeton physicists were in the process of designing an experiment to measure the relic radiation that they recognized was a crucial implication of the Big Bang theory when they discovered they had been scooped—by the Bell Lab scientists who hadn't yet realized what they had found. Luckily for Penzias and Wilson, the Massachusetts Institute of Technology astronomer Bernie Burke, whom Robert Wilson described to me as his personal early Internet, knew about both the Princeton research and also the mysterious finding of Penzias and Wilson. Burke put two and two together and clinched the connection by bringing the relevant players into contact. After consulting with the theoretical physicist, Robert Dicke, Penzias and Wilson recognized the import and value of what they had discovered. Along with the much earlier discovery of the Hubble expansion, this discovery of the background radiation, which subsequently earned the two Bell Lab physicists the Nobel Prize in 1978, clinched the Big Bang theory of a cooling, expanding Universe.

This was a lovely example of science in action. The research was done for a specific scientific purpose but had ancillary technological and scientific benefits. The astronomers weren't looking for what they found, but because they were extremely technologically and scientifically skilled, they didn't dismiss their finding. The research—while looking for relatively small discoveries—resulted in a discovery with tremendously deep implications, which they found because others were simultaneously thinking about the big picture. The discovery by the Bell Lab scientists was accidental, but it forever changed the science of cosmology.

Furthermore, within a few decades of its discovery, this radiation also helped advance major new insights into cosmology. In a spectacular achievement, detailed measurements of this radiation helped verify the predictions of *cosmological inflation*—in which an explosive stage of expansion occurs very early on.

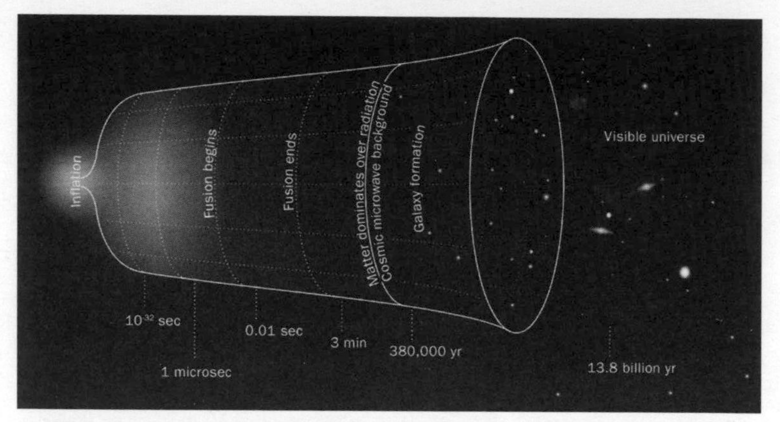

[FIGURE 5] History of the Universe with inflation and Big Bang evolution, including the formation of nuclei, structure beginning to form, the cosmic microwave background radiation imprinted on the sky, and the modern Universe—in which galaxies and galaxy clusters have been established.

COSMOLOGICAL INFLATION

Many scientific breakthroughs have emerged from an underlying debate about whether change happens gradually, or suddenly, or even—as with our initial ignorance of the Universe's expansion—if change happens at all. Although people frequently neglect the relevance of this important factor, accounting for the rate of change in today's world can be very useful when considering the consequences of technology, for example, or when evaluating environmental transformations.

Debates about the pace of change underscored many of the central nineteenth-century conflicts over Darwinian evolution too. As we will see in Chapter 11, the debates contrasted gradualism as espoused by Charles Lyell, in the case of geology, and his acolyte Charles Darwin with the arguments in favor of sudden geological changes posed by the Frenchman Georges Cuvier. Cuvier also recognized another kind of radical change, suggesting, controversially, that not only do new species emerge, as Darwin had so notably demonstrated, but that they also disappear through extinction.

Debates about the pace of change were also central to our understanding of the development of the cosmos. With the Universe, the first surprise was that it evolves at all. When the Big Bang theory was first proposed in the early twentieth century, its implications were very different from those of the theologically favored static Universe, which was what most people had accepted at the time. But another, later surprise was the recognition that very early on, our Universe underwent a phase of explosive expansion—cosmological inflation. As with life on Earth, both gradual and catastrophic processes played a role in the Universe's history. For the Universe, the "catastrophe" was inflation. And by catastrophe, I mean only that this phase occurred suddenly and rapidly. Inflation destroyed the contents of the Universe that had been there initially, but it also created the matter that filled our Universe when the explosive phase came to an end.

The history presented so far is the standard Big Bang theory of an expanding, cooling, aging Universe. It is remarkably successful, but it's not the whole story. Cosmological inflation occurred before the standard Big Bang evolution took over. Even though I cannot tell you what happened at the very beginning of the Universe, I can say with reasonable certainty that at some time very early on in its evolution—perhaps as early as 10^{-36} seconds in—this sensational event called inflation took place. (See Figure 5.) During inflation, the Universe expanded far more rapidly than it did during standard Big Bang evolution—most likely exponentially—so that the Universe kept multiplying in size for the duration of this inflationary phase. Exponential expansion means, for example, that when the Universe was sixty times older than when inflation began, the Universe would have increased by more than a trillion trillion–fold, whereas without inflation, the Universe's size would have increased only by a factor of eight.

Once inflation ended—also only a fraction of a second into the Universe's evolution—it left behind a large, smooth, flat homogeneous Universe whose later evolution is predicted by the traditional Big Bang theory. The inflationary explosion was in a sense the "bang" that started the cosmological evolution toward the smoother, slower evolution that was just described. Inflation diluted away the initial matter and radiation as the rapid cooling sent the temperature very close to zero. Hot matter was reintroduced only when inflation ended and the energy driving inflation was converted to a tremendous number of elementary particles. The conventional, slower expansion took over when inflation ended. From this stage forward, the old Big Bang cosmology applies.

The physicist Alan Guth developed the theory of inflation because the Big Bang theory—successful as it is—left several issues unresolved. Why, if the Universe grew from an infinitesimally sized region, is there so much stuff contained within? And why has the Universe been so long-lived? Based on the theory of gravity, you might have expected a Universe containing so much stuff to have expanded away into nothingness or to have collapsed very quickly. Yet despite the enormous amount of matter and energy it contains, the three infinite spatial dimensions of the Universe are very nearly flat and the Universe's evolution has been sufficiently slow for us to celebrate its 13.8 billion years of existence.

One further major omission in the original Big Bang cosmology was an explanation for why the Universe is so uniform. When the cosmic radiation we now observe was emitted, the Universe was only about one-thousandth its current size, meaning that the distance light could have traveled was far smaller. Yet when observers view the radiation emitted from different regions of the sky from this time, the radiation appears to be identical, meaning that the deviations in temperature and density are minuscule. This is puzzling because according to the original Big Bang scenario, the age of the Universe at the time when the cosmic radiation decoupled from charged matter was too low for light to have had enough time to travel even one percent of the way across the sky. That is to say, if you go back in time and ask whether the radiation that ends up in these separated patches of the sky could have ever sent or received any signals between them, the answer would be no. But if the separate regions never communicated with each other, why would they look the same? It would be as if you and a thousand strangers from different places with different stores and different magazines as inspiration entered a theater dressed identically. If you never had any contact with each other or with shared media outlets, it would be a remarkable coincidence for you to all end up dressed alike. The sky's uniformity is even more remarkable since the uniformity applies at a precision of one in 10,000. And it looks like the Universe began with more than 100,000 regions that didn't communicate with each other.

The idea that Guth proposed in 1980 seemed very attractive in light of these deficiencies. He suggested an early epoch during which the Universe expanded extraordinarily rapidly. Whereas in the standard Big Bang scenario, the Universe grew calmly and steadily, in the inflationary epoch, the Universe underwent a phase of explosive expansion. According to the theory of cosmological inflation, the very early Universe grew from a tiny region to an exponentially larger region in an extremely short period of time. The size of a region that a light ray could have crossed might have increased by a factor of a trillion trillion. Depending on when inflation began and how long it lasted, the original region a light ray could cross might have begun as 10^{-29} meters in size but expanded during inflation to be at least about a millimeter big—a little bigger than a piece of sand. With inflation, you do in some sense have the Universe in a grain of sand—or at least in the size of a grain of sand, as William Blake would have you believe, if you measure the size of the Universe as the observable region at that time.

The inflationary Universe's extremely rapid expansion explains the Universe's enormity, uniformity, and flatness. The Universe is enormous because it grew exponentially—in very little time it became very big. An exponentially expanding Universe covers far more territory than one expanding at the far slower rate of the original Big Bang scenario. The Universe is uniform because the enormous expansion during inflation smoothed out the wrinkles in the spacetime fabric, much as stretching out your jacket sleeve eliminates the creases in its fabric. With the inflationary Universe, a single very small region in which everything was close enough to communicate via radiation grew into the Universe we see today.

Inflation also explains flatness. From a dynamical perspective, the flatness of the Universe means that the density in the Universe as a whole is at the borderline where it can last a very long time. Any larger energy density would have made for positive curvature of space—the kind of curvature that a sphere has—which would have made the Universe quickly collapse. Any less density would have caused the Universe to expand so rapidly that structure would never have coalesced and formed. Technically, I am overstating slightly. With a very tiny amount of curvature, the Universe could have lasted as long as it has. But that curvature would have had to have been mysteriously small without inflation to justify its value.

In an inflationary scenario, the Universe is currently so large and flat because it grew so much early on. Imagine that you could blow up a balloon to be as large as you wanted. If you focus on some particular region of the balloon, you would see that it became flatter as the balloon grew larger. Similarly, people originally thought the Earth was flat because they saw only a small region of the surface of a much larger sphere. The same thing is true for the Universe. It flattened as it expanded. The difference is that it expanded by a factor exceeding a trillion trillion.

The extreme flatness of the Universe was the chief confirmation of inflation. This might not come as a surprise, since flatness was after all one of the problems inflation was supposed to address. But at the time inflation was conceived, it was known that the Universe was flatter than naive expectations would suggest, but with nowhere near the precision required to test inflation's extreme prediction. The Universe has now been measured to be flat at the level of a percent. Had this not been true, inflation would have been ruled out.

When I was a graduate student in the 1980s, inflation was considered an interesting idea, but not one that most particle physicists took very seriously. From a particle physics point of view, the circumstances required for a long-lived exponential expansion seemed extremely unlikely. In fact they still do. Inflation was supposed to address the naturalness of the initial conditions for the expansion of the Universe. But if inflation itself is unnatural, the problem isn't really solved. The question of how inflation occurred—its underlying physics model—remains a matter of speculation. The model-building issues that plagued us in the 1980s are still a concern. On the other hand, people like Andrei Linde, a Russian-born physicist now at Stanford, who was one of the first to work on inflation, thought it had to be correct even when the idea was first proposed simply because no one had found any other solutions to the puzzles of size, flatness, and homogeneity, which inflation was able to solve in one fell swoop.

In light of recent detailed measurements of the cosmic microwave background radiation, most physicists now agree. Despite the fact that we have yet to determine the theoretical underpinnings of inflation, and that inflation happened long ago, it leads to testable predictions, which have convinced most of us that inflation, or something very similar to inflation, has occurred. The most precise of these observations concerned details about the 2.73 degree background radiation that Penzias and Wilson had discovered. NASA's Cosmic Background Explorer (COBE) measured this same radiation, but more comprehensively and over a large range of frequencies—establishing its extremely high degree of uniformity across the sky.

But the most spectacular COBE discovery—one that won over almost all inflationary skeptics—was that the early Universe was not exactly uniform. Overall, inflation made the Universe extremely homogeneous. But inflation also introduced very tiny *inhomogeneities*—deviations from perfect uniformity. Quantum mechanics tells us that the exact time at which inflation ends is uncertain, which means that it ended at slightly different times in different regions of the sky. These tiny quantum effects were imprinted in the radiation as small deviations from perfect uniformity. Though far smaller, they are like the perturbations that rise in the water when you drop a pebble into a pond.

In what is certainly among the most mindblowing discoveries of the last few decades, COBE discovered the quantum fluctuations that were generated when the Universe was roughly the size of a grain of sand, and which are ultimately the origin of you, me, galaxies, and all the structure in the Universe. These initial cosmological inhomogeneities were generated when inflation was ending. They started on minuscule-length scales but they were stretched by the expansion of the Universe to sizes where they could seed galaxies and all other measurable structure, as the following chapter will explain.

Once the discovery of these density perturbations—as these small deviations in temperature and matter density are known—was made, it was only a matter of time before they were investigated in detail. Beginning in 2001, the Wilkinson Microwave Anisotropy Probe (WMAP) measured density perturbations with even more accuracy and on smaller angular scales. WMAP, along with telescopes at the South Pole, observed the ripples—perturbations—in the density of the radiation that encapsulates the complexity that had just begun to be created. The details of these measurements confirmed the Universe's flatness, determined the total amount of dark matter, and verified the predictions of an early exponential expansion. Indeed, one of the most fabulous results from WMAP was its experimental confirmation of the inflationary paradigm.

The European Space Agency launched its own satellite—the Planck mission—in May 2009 to study the perturbations in even more exquisite detail. Indeed, the satellite's results have improved the precision with which most cosmological quantities are known and helped solidify our knowledge of the early Universe. One of the Planck satellite's most important accomplishments was that it pinned down one further quantity that hints at the dynamics that drove the inflationary expansion. Just as the Universe is mostly homogeneous, with small perturbations that violate this homogeneity, the amplitude of perturbations in the sky is mostly independent of their spatial extent, but exhibits a small dependence on scale. The dependence on scale reflects the changing energy density of the Universe at the time inflation ended. In an impressive confirmation of inflationary dynamics, WMAP and the more precise Planck satellite measured that scale-dependence, determining that an early stage of rapid expansion gradually came to an end, and measuring a value that constrains inflationary dynamics.

Although our understanding is far from complete, cosmologists have now established that inflation and the subsequent Big Bang expansion are part of our Universe's history. We can establish these theories in detail because the early Universe, with its high degree of uniformity, is relatively easy to study. Equations can be solved and the data can be readily evaluated.

However, billions of years ago when structure formation took place, the Universe changed from a relatively simple system to a far more complex one, so cosmology faces greater challenges when addressing the Universe's later evolution.. The distribution of the contents of the Universe became more difficult to predict and interpret as structures such as stars, galaxies, and galaxy clusters were formed.

Nonetheless, a great deal of information is buried in this ever-evolving structure of the Universe—which observations, models, and computer power should ultimately reveal. As we'll see in the later part of the book, measuring and predicting this structure promises to teach us quite a lot—including the relevance of dark matter to our world. But for now, let's explore how this structure came about in the first place.

5

A GALAXY IS BORN

You might recall my dinner conversation in Munich in which Massimo, the branding expert, objected to the name "dark matter." At that same dinner, Matt, another conference participant to whom Massimo introduced me, inquired about people's potential for harnessing this elusive matter's power. This was an understandable question for the game designer that he was. A screenwriter friend asked the same thing shortly after—again, not surprising, in light of her science fiction predilections.

But this inquiry represents very wishful thinking, which I'll once again attribute to the poor choice of name. Dark matter is neither an ominous—nor a munificent—source of strategic power in our local neighborhood. Given the extraordinary feebleness with which known matter could possibly influence dark matter, no one can collect it in a basement or garage. With our hands and tools made from ordinary matter, we can't make dark matter missiles and we can't make dark matter traps. Finding dark matter is difficult enough. Harnessing it would be another thing altogether. Even if we could find a way to contain dark matter, it wouldn't affect us in any noticeable way since it interacts only through gravity or through forces that have so far been too weak to detect—even by very sensitive searches. In the absence of enormous astronomical-sized objects, dark matter's influence on Earth is too small to care about. This is also why it's so hard to find.

But the large amount of dark matter collected by the Universe as a whole is another thing altogether. The enormous amount of dark matter that is spread throughout the Universe collapsed and collected to create galaxy clusters and galaxies, and these in turn allowed for the formation of stars. Although dark matter hasn't directly influenced people or laboratory experiments (yet) in any recognizable fashion, its gravitational influence was critical to the formation of structure in the Universe. And because of the large amounts concentrated in these enormous collapsed regions where matter is situated, dark matter continues to influence the motion of stars and the trajectories of galaxies today. As we will soon see, a less conventional type of dark matter that collapsed to become even denser could conceivably be influencing the trajectory of the Solar System too. So even though people can't harness the force of dark matter, the far more powerful Universe can. This chapter will explain dark matter's critical role in the Universe's evolution, and in the formation of galaxies during its known, finite lifetime.

THE EGG AND THE CHICKEN

The theory of structure formation tells us how stars and galaxies developed from the extremely—but not completely—boring, uniform sky that was the final legacy of inflation. This consistent picture of structure formation, like so much presented in this book, is a relatively recent advance. But this theory is now firmly grounded in cosmological developments, such as the Big Bang theory supplemented by inflation, and in better-measured ingredients such as dark matter. These underpinnings let us explain how the hot, disorderly, undifferentiated region that constituted the incipient Universe developed into the galaxies and stars that we see today.

Initially the Universe was hot, dense, and mostly *uniform*—the same at every point in space. It was also *isotropic*—which means it is the same in all directions. Particles interacted, appeared, and disappeared, but the density and behavior of the particles were everywhere the same. This is of course a very different image from the one you see when looking at pictures of the Universe, or when you simply shift your gaze upward to admire the beauty of the night sky.

The Universe is no longer uniform. Galaxies, galaxy clusters, as well as stars, punctuate the expanse of space, staking their uneven distribution throughout the firmament. Such structures are at the heart of everything in our world, which couldn't have been created without the dense stellar systems that were essential to the formation of the heavy elements and all the amazing stuff, including life, which developed in at least one such concentrated stellar environment.

The visible structure of the Universe lies in gas and in *stellar systems*. These congregations of stars come in a wide variety of sizes and in several different shapes. Binary stars, with one star rotating around another, constitute a stellar system, as do galaxies, which range in size from one hundred thousand to a trillion stars. Clusters of galaxies with a thousand times as many stars as that are stellar systems too.

To get some appreciation of the types of objects involved, let's consider typical masses and sizes for the objects that our cosmos contains. Astronomical sizes are generally measured in parsecs or light-years, whereas astronomical masses are usually measured in terms of solar masses—how many Suns it would take to yield the equivalent mass. Galaxies range in size from smaller than dwarf galaxies of about 10 million solar masses to the largest galaxies with about 100 trillion solar masses. The Milky Way galaxy sports a smaller, more typical size of about a trillion solar masses—this value represents its total mass, including the dominant dark matter component. Most galaxies have sizes between a few thousand and a few hundred thousand light-years in diameter. Galaxy clusters, on the other hand, contain between 100 trillion and 1,000 trillion solar masses, with diameters typically on the order of 5 to 50 million light-years. Clusters contain up to about a thousand galaxies, whereas superclusters can contain 10 times that number.

However, although these objects exist today, the early Universe didn't contain them. The early Universe was extremely dense, so it didn't yet contain stars or galaxies, which have far lower densities. Stellar systems could form only after the Universe had cooled to a temperature at which it had lower average density than that of the objects that eventually formed. The formation of structure also had to wait until matter in the Universe carried more energy than radiation. Note that I am using the cosmological definition of radiation, which is anything, including particles such as photons that travel at or near the speed of light. In the hot early Universe, almost everything satisfied this criterion because the temperature was so high, allowing radiation to dominate the Universe's energy.

As the Universe expanded, radiation and matter both diluted, and so too did their energy densities. Since the energy in radiation, which redshifts to lower energy, dies away more quickly, matter—after waiting 100,000 years for its turn in the spotlight—eventually came to dominate the Universe's energy. At this landmark epoch, matter overtook radiation as the leading contributor to the Universe's energy.

A good starting point for following how structure first grew is about this time, 100,000 years into the Universe's evolution, when matter began to dominate. This epoch is relatively late compared to the time when perturbations first began to grow, but it is not all that long before the background radiation we observe was imprinted. Matter-domination was significant for cosmology because slowly moving matter carries far less pressure than radiation, and therefore influences the Universe's expansion differently. When matter takes over, the expansion rate of the Universe changes. But more important for structure formation, small, compact structures could then begin to grow. Radiation, which moves at or very near the speed of light, doesn't slow down enough to get trapped into small gravitationally bound systems. Radiation washes out perturbations, much as wind erases ripples of sand imprinted on a beach. Matter, on the other hand, can slow down and clump together. Only slowly-moving matter collapses sufficiently to form structure.

This is why cosmologists sometimes say that dark matter is *cold*, which means that it's not hot and relativistic and doesn't act like radiation.

After matter dominated the energy density of the Universe, density perturbations—regions that are slightly denser or less dense than others, which are created when inflation ended—precipitated the collapse of matter that seeded the growth of structure. These perturbations subsequently grew and transformed the initially homogeneous Universe into what would ultimately be amplified into differentiated regions in the sky. The tiny variations in density at the level of less than 1 in 10,000 sufficed to create structure from a nearly homogeneous Universe because it is flat, which means that it has the critical energy density that straddles the border between rapid collapse and rapid expansion. Critical density creates the sweet spot at which the Universe expands slowly and lasts long enough for structure to form. In this delicately determined environment, even small density perturbations cause regions of matter to collapse, thereby initiating the formation of structure.

Two competing forces contributed as this collapse into structure began. Gravity pulled matter in, while radiation—though not the dominant type of energy—pushed it out. The threshold beyond which this balance would be destroyed is known as the *Jeans mass*. The gas inside the region in which the outward radiation pressure doesn't balance the inward gravitational pull collapses, and the matter and the objects that grew from its attractive potential became the seeds of luminous galaxies and the formation of stars.

The regions with greater density exerted more gravitational attraction than those with lower density, thereby creating increasingly dense regions and further depleting the surrounding, already diffuser, domains. The Universe grew lumpier as the (matter-)rich regions got richer and the(matter-)poor domains got poorer. This aggregation of matter continued—creating gravitationally bound objects—and matter continued collapsing in a positive feedback process. Stars, galaxies, and clusters of galaxies all were created at this time by gravity's effects on the initial tiny quantum mechanical fluctuations that were created at inflation's end.

Because of its immunity to radiation and because of its greater abundance, the majority of the matter creating the attractive potential wells that pulled matter in to collapse initially was dark matter—not ordinary matter. Though we see stars and galaxies because of emitted light, dark matter is what initially attracted the visible matter to these denser regions where galaxies and subsequently stars could emerge. When a sufficiently large region collapsed, dark matter formed a roughly spherical halo inside of which the gas of ordinary matter could cool, condense into the center, and eventually fragment into stars.

Regions collapsed sooner in the presence of dark matter than would have been possible with only ordinary matter because the greater total matter energy density allowed matter to dominate over radiation sooner. But dark matter was also important because electromagnetic radiation initially prevented ordinary matter from developing structure on scales smaller than about a hundred times the size of a galaxy. Only by hitchhiking with dark matter did galaxy-sized objects and the seeds of stars in our Universe have time to form. Without dark matter initiating the collapse, stars wouldn't have reached their current population and distribution.

So it's dark matter that started the collapse into structure. Not only is there more of it, but because dark matter is essentially immune to the influence of light, electromagnetic radiation could not drive it apart the way it can with ordinary matter. Dark matter thereby established the fluctuations in the matter distribution that ordinary matter responded to when radiation decoupled. Dark matter effectively gave ordinary matter a head start—paving the way for the formation of galaxies and stellar systems. Because it is immune to radiation, it could collapse even when ordinary matter could not, forming a substrate in which protons and electrons could be shepherded into collapsing regions.

This simultaneous collapse of dark matter and ordinary matter into visible objects such as galaxies and stars is important for structure formation, and also for observations. Even though we directly see only ordinary matter, we can be pretty confident that dark matter and ordinary matter exist in the same galaxies. Since ordinary matter relied on dark matter to seed structure—ordinary matter, which came along for the ride, resides mostly in structures that contain a significant amount of dark matter too. So, in a sense, looking under the lamppost for dark matter is appropriate.

It is also worth noting that dark matter continues to play an important role today. Not only does it contribute to the gravitational attraction that keeps stars from flying away, but it also attracts back into galaxies some of the matter that is ejected by supernovae. Dark matter thereby helps retain heavy elements that are essential to further star formation and ultimately to life itself.

However, although physicists can predict early structure formation based on theory, no observer can currently witness in detail the Universe's transition during initial structure formation. Telescopes detect light that was emitted in more recent times, and even let us examine the earliest galaxies formed billions of years ago. The observable cosmic microwave background radiation, on the other hand, comes to us from a time when the universe was full of radiation—but when collapsed gravitationally bound objects had yet to form. The background radiation imprinted early density fluctuations for posterity 380,000 years into the universe's evolution, but it would be about another half billion years before any stars or galaxies would exist and emit observable light.

The intermediate time after recombination—when neutral atoms formed and the cosmic microwave radiation was imprinted—and before luminous objects came about was a very dark epoch that is not accessible to current observational instruments. Objects didn't emit light because stars hadn't yet formed, yet the microwave background radiation that earlier on had interacted with the ubiquitous electrically charged matter no longer lit up the sky. This time is invisible to conventional telescopes (See Figure 6.) Yet this is exactly the epoch when the primordial soup transformed into the progenitor structures of the rich and complex Universe we can now observe.

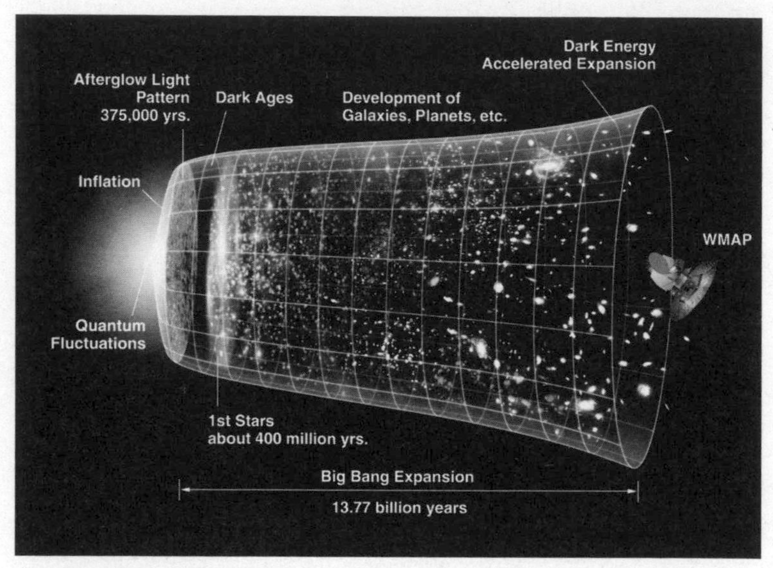

[FIGURE 6] After the time we observe through the cosmic microwave background radiation, a dark era ensues in which structure forms, which is followed by the emergence (and dissolution) of the first stars, the subsequent formation of galaxies and other structures, and dark energy coming to dominate the Universe's expansion. (NASA)

The Harvard astrophysicist Avi Loeb compares our inability with current technology to watch the earliest stars form to our inability to witness the formation of a chicken from an egg. An egg contains a gloppy, souplike structure. But let a hen sit on it long enough and out of that egg will emerge a functioning chick that will continue to develop into a fully grown chicken. The yolk and albumen that we know about from broken eggs look nothing like what comes out, but they contain all the seeds of the chick that will emerge. But the transition happens inside the shell, so no one can see what is happening without special tools.

Similarly, we will need new technology to witness the earliest formation of structure. No one can now see the dark period in the universe's evolution—though proposals are in the works. Yet we know that density perturbations, like an egg, contain the seeds of later structure. But, unlike the chicken/egg quandary, we do know what came first.

HIERARCHICAL STRUCTURE

The above picture of structure formation—based on the process of individual perturbations that seed individual galaxies that then evolve independently of each other—contains much of the relevant physics of collapse. Further investigation shows that giant stars formed first—but they either quickly exploded into supernovae, releasing the first heavier elements into the Universe, or they collapsed into black holes. These heavy elements played an important role in the Universe's subsequent development. Only after metals—which is what astronomers call heavy elements—were present could the smaller stars (like our Sun) form in cooler, denser regions, and the structure we now observe be created.

But before those stars could form, galaxies had to emerge. Indeed, galaxies were the first complex structures to exist. Galaxies—each seemingly self-contained but, as we will soon see, all connected—were in many respects the building blocks of the Universe. Once formed, galaxies could merge into larger structures, like galaxy clusters. And after sufficient collapse, stars can form inside the densest regions. But the formation of the structure that we see today began with galaxies.

However, this picture of galaxies forming individually is a simplification. In reality, galaxies are not isolated island universes as this picture would have you believe. Encounters and mergers with other galaxies are critical to their development. Galaxy formation is hierarchical, with smaller galaxies forming first and larger structure following. Even galaxies that appear to be isolated are surrounded by larger dark halos that are contiguous with the halos of other galaxies.

Because the galaxies occupy a reasonably large fraction of space—about 1 in 1,000—galaxies collide far more often than stars, which occupy a volume more like 1 in 10 million trillion. Through mergers and other gravitational interactions, galaxies continue to influence each other. Galaxies further evolve as they continue to attract gas, stars, and dark matter into their fold.

Armed with this further knowledge, let's reexamine what happens in structure formation. To better understand the process, the rich-get-richer and poor-get-poorer analogy is remarkably apt. As is discussed in today's world with increasing frequency and urgency, the poor not only become poorer. They become more numerous too. In fact, some hotly debated apocalyptic scenarios for humanity that I sometimes hear discussed predict that the rich will become crowded into small domains, edged out to the margins by the spreading of the far more populous poor societies. In this not very appealing scenario, the rich will reside on the outskirts of towns, like I saw when I visited the white suburban neighborhoods in the outer regions of Durban, South Africa. But then—continuing the analogy—neighboring towns will experience similar phenomena too. Once sufficiently spread out, the neighborhoods will collide, leaving the rich only at the intersections. The rich, segregated population might then invest in businesses and security systems, but all this development and rapid growth will be left to the nodes where the privileged classes of society crossed.

Though not an attractive picture for society, this is remarkably similar to the way that structure formation in the Universe proceeds. Underdense regions expand more rapidly than the Universe as a whole, whereas overdense regions expand more slowly. As a result, the underdense regions crowd out the overdense ones, leaving them only on the margins of the originally expanding low-density regions. The diffuser regions get depleted and evolve into voids, and as they do so they grow—shepherding matter into high-density sheets at the boundary.

When such sheets intersect, filaments of high-density regions form. The gravitational attraction from these regions collects all the remaining "wealth" of matter. These increasing amounts of matter get confined to a cosmic web of thin, dense sheets enclosing voids. This cosmic web becomes a network of filaments wherein the densest matter lies at the nodes where filaments intersect. So rather than simple spherical collapse, material falls first along sheets into filaments that intersect to form nodes. (See Figure 7.) These nodes then seed the formation of galaxies. And this process continues over time. Structure forms and patterns repeat on increasingly larger scales. This yields a hierarchical bottom-up model where smaller structures form before larger ones, in which small galaxies formed first.

[FIGURE 7] Simulation of the "cosmic web" of matter: Filaments of dark matter that intersect at nodes enclose dark, relatively empty voids. Galaxy clusters indicated by the very bright regions, form at the nodes. (Image of the projected dark matter density through a slice 18 Mpc thick with side length of 179 Mpc created by Benedikt Diemer and Philip Mansfield, using the 2012 visualization algorithm of Kaehler, Hahn and Abel.)

Numerical simulations confirm these predictions on the largest scales, with dark matter correctly accounting for the density and shape of structure in the Universe. Smaller-scale discrepancies might be clues to further refinements of this theory, but we will leave for later a discussion of these less well-established predictions and observations and the models that might resolve them.

Because ordinary matter and dark matter collapse in sync, radiation from galaxies traces dark matter–heavy regions too. In the same way that ambient light over the globe maps its cities, the brightest regions in the Universe map the densest galactic regions with the largest numbers of stars. The light traces the overall mass density, much as the light map of the world traces the population density.

However, we should keep in mind that as with light, the proportion of what we see to the actual population can vary. The ratio of dark to light matter depends on whether the object is a dwarf galaxy, a galaxy, or a galaxy cluster, for example. Nonetheless, even with varying ratios, where there is light, there is dark too. This is a valuable observational tool in verifying the theory of structure formation.

OUR 'HOOD

Before closing this chapter—and the first part of this book—let's now turn to the distribution and influence of ordinary matter within the galaxy we know best, the Milky Way, and our favorite star within it, which is the Sun. Our galaxy is named for the band of milky white light that is visible in the sky on a clear, dry night. That light arises from the cumulative light of the myriad faint disk stars that lie within the Milky Way plane. Despite the suggestive wrapping on the dark chocolate version, the Milky Way candy bar (which I quite like and of which I have eaten too many) is named instead for the malted milk shake—whose delicious flavor the industrially produced treat is supposed to share.

The Milky Way galaxy is in a group of galaxies knows as the Local Group, which is a gravitationally bound system of galaxies whose density is higher than average. The Milky Way and the Andromeda galaxy, also known as M31, dominate this group's mass, but dozens of smaller galaxies belong to the group too—mostly satellites of the two bigger ones. The gravitational binding force of the Local Group prevents the Milky Way and Andromeda from receding from each other with the Hubble expansion. Their paths are actually converging and in about four billion years they will collide and merge.

THE MILKY WAY

The Milky Way has a disk of gas and stars that extends roughly 130,000 light-years across and about 2,000 light-years in the vertical direction, with this "flattish-land" structure yielding its distinctive shape. The disk contains stars, as well as hydrogen gas and small, solid particulate dust in what is known as the interstellar medium, which in total has a mass about a tenth the net mass of the stars. We don't actually see the brighter concentration of light near the center of the galaxy where most stars reside since the interstellar dust obscures the light. Astronomers do, however, see the galaxy center in the infrared, since dust doesn't absorb this lower-frequency light. The center of the Milky Way also contains a black hole of about four million solar masses—known sometimes as Sagittarius A*.

The black hole in the center and dark matter are entirely different things. However, dark matter does exist in a large spherical halo—about 650,000 light-years wide. This largest component of our galaxy in terms of size and mass carries about a trillion solar masses in a roughly spherical region that encompasses the Milky Way disk. As with all galaxies, dark matter condensed first and attracted the ordinary matter that makes up what we see. (See Figure 8.)

But I have yet to describe how and why a disk would form and this will be important to the dark matter disk idea and its consequences for meteoroids, which I will discuss in detail later on. Ordinary matter is interesting in that it can have a very different distribution within a galaxy than dark matter. Dark matter forms a diffuse spherical halo, whereas ordinary matter can collapse into a disk, such as the familiar disk of stars of the Milky Way plane.

Ordinary matter's interaction with electromagnetic radiation is responsible for this collapse. An important distinction between ordinary matter and dark matter is that ordinary matter can radiate. Without

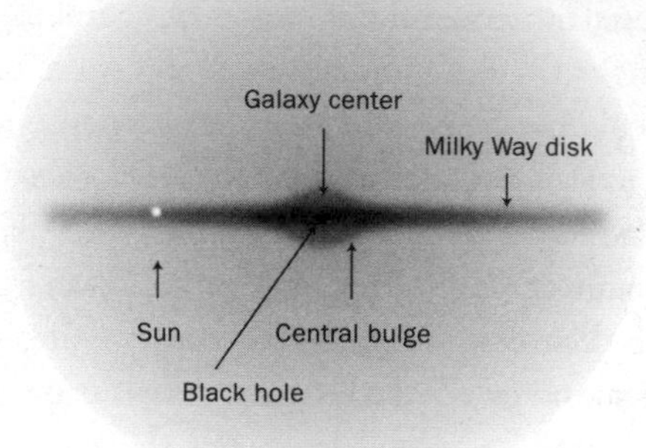

[FIGURE 8] Milky Way disk with its central bulge, black hole, and a surrounding dark matter halo. Also indicated is the position of the Sun (whose size is not to scale).

the radiation that leads to cooling, ordinary matter would remain as diffuse as dark matter. Actually, it would be even less dense since its energy budget is only about one fifth as large. However, the interactions of ordinary matter with photons allow it to dissipate energy and cool so that it can collapse into a more concentrated region—namely a disk. The energy loss via emission of photons is akin to evaporation, in which the vaporization of water carries energy away from your skin. But unlike dissipating matter, you usually don't collapse when you sweat and cool down. However, because ordinary matter can dissipate energy, gas collapses and concentrates in a smaller collapsed region, in which it rises to higher density than dark matter.

The reason ordinary matter lies in a disk and not a small ball is the matter's net rotation, which it inherited from the gas clouds that acquired angular momentum (momentum of rotation) in their formation. Cooling lowers resistance to collapse in one direction, but collapse in the two others is prevented or at least lessened by the centrifugal force of the rotation of the gas it contains. Without friction or some other force acting on it, a marble that you set in motion around a circular track will keep rolling forever. Similarly, once matter is rotating, it will keep its angular momentum until some torque acts on it or it can dissipate angular momentum along with energy.

Because angular momentum is conserved, gaseous regions cannot collapse as efficiently in the radial direction (as defined by rotation) as in the vertical one. Though matter might collapse in the direction parallel to the axis of rotation, it won't collapse in the radial direction unless angular momentum is somehow removed. This differential collapse is what gives rise to the relatively flat disk of the Milky Way, which we observe stretching across the sky. It is also what gives rise to the disks of most spiral galaxies.

THE SUN AND THE SOLAR SYSTEM

The galaxy's net mass is dominated by dark matter, but ordinary matter, which is concentrated in the Milky Way disk, dominates the physical processes in the Milky Way plane. Although ordinary matter has only a limited role in initiating structure, with its elevated density and its nuclear and electromagnetic interactions, ordinary matter is critical to many important physical processes—including star formation.

Stars are hot, dense gravitationally bound balls of gas that are fueled by nuclear fusion. They get created in the dense gaseous regions of a galaxy. As the gas in the disk orbits around the galaxy's center, it breaks into clouds that are denser regions that can collapse further. Stars were formed out of the gas that collapsed to very high densities inside those haloes.

One of those gaseous balls, our Sun, began 4.56 billion years ago as an energetic system in which gravity, gas pressure, magnetic fields, and rotation all played a role. Meteorites have been found that include material almost as old as the Solar System, and many museums have examples. The Sun is located very close to the midplane of the Milky Way disk and is at a radius of about 27,000 light-years—farther out radially than at least three-quarters of the other stars.

Like the other hundred billion stars in the Milky Way disk, the Sun circles around the galaxy at a speed of about 220 kilometers per second. At this speed, it takes about 240 million years for the Sun to orbit the galactic center. Since the galactic plane is less than 10 billion years old, the stars in the plane have made less than 50 revolutions in that time. It's enough time for the system to homogenize some gross features, but really it's not all that many trips around.

The Solar System and its formation are among the many scientific subjects where knowledge has blossomed in the last several decades. As with most stars, the Sun and the Solar System emerged from a giant molecular cloud of gas. Before the Sun was born, everything in the solar vicinity moved rapidly and collisions frequently occurred. After about a hundred thousand years, the system collapsed into a *protostar,* in which nuclear fusion wasn't yet occurring, and a *protoplanetary* disk, which is what would ultimately transmute into planets and other objects in the Solar System. About 50 million years later, hydrogen began to fuse and what we think of as our Sun came into being. The Sun gobbled up most of the mass from the nebular cloud, but some material remained to collect in a disk around the Sun from which the planets and other Solar System objects, such as comets and asteroids, would emerge. Once the energy produced by the Sun resisted gravitational contraction, the Solar System was born.

What really caught me and my collaborators by surprise was the critical role that molecules and heavy elements play in cooling the gas sufficiently to allow for the formation of most stars. Heavy elements aren't only important for nuclear burning. They are also essential for allowing matter to cool by scattering to the point where burning is even an option. Solar-sized star formation requires extremely cool temperatures—tens of kelvins. Overly high temperature gas never gets sufficiently concentrated to ignite nuclear burning. In yet another amazing connection between basic fundamental processes and the nature of the Universe, without the heavy elements and molecular cooling that ordinary matter experiences, the gas that created the Sun would never have sufficiently cooled down.

Only after beginning my more recent research—which focuses on the details of astronomical systems more than my earlier particle physics work—did I truly appreciate the beauty and coherence of the Universe's dynamical systems. Galaxies are formed, stars get created, and the heavy elements created by those stars and the gas they eject contribute to further star formation. Despite its appearance on human time scales, the Universe and everything inside it is far from static. Not only do stars evolve, but galaxies do as well.

The next part of the book focuses on the Solar System, and discusses asteroids, comets, impacts, as well as life's emergence and disappearance. We will see that the same pattern of interactions and changes is true too in our more immediate environment.

PART II

AN ACTIVE SOLAR SYSTEM

6

METEOROIDS, METEORS, AND METEORITES

I was delighted when visiting the desert near Grand Junction, Colorado, to have someone lend me a pair of night-vision goggles. These specially designed glasses, which are so powerful that laws currently forbid exporting them outside the United States, amplify light so that a great deal that is ordinarily too dim for human eyes becomes sufficiently illuminated to see. The military uses them to look for enemy combatants and mountain residents employ them to find nocturnal animals.

Interested in neither of these uses, I took advantage of the opportunity to look up at the sky, where I could spot objects so faint I would never have noticed them without assistance. By far the most striking thing I noticed in the clear, dry skies above was the frequency of "shooting stars"—small meteoroids burning up in the atmosphere. Within the span of a couple of minutes, perhaps five or ten had zoomed across my field of view. I was lucky since I had looked up during a meteor shower, so the light streaks passed by even more frequently than they ordinarily would. But even without meteor showers enhancing the rate, grains of sand burn up in the atmosphere all the time.

The meteors created by those grains of dust are nothing if not exciting. Magnificent light displays overhead, emanating from dust or pebbles flying by in space radiate romance and mystery. That is, when they don't trigger thoughts of destruction. No one wants to be hit by a high-speed rock, no matter how small. And we certainly don't want a big rock to hit the Earth. Happily, although on very rare occasions objects sizable enough to do damage have hit or come close, most of the stuff that makes it anywhere near us is not cause for concern. About 50 tons of extraterrestrial material enters the Earth's atmosphere every day, carried by millions of small meteoroids. And none of us are affected in any noticeable way.

The first part of this book focused on dark matter and the Universe as a whole, with a brief shout-out to the Milky Way and the Solar System at the end. This part concentrates on our Solar System, and especially to those objects that could be relevant in the presence of a dark disk. It explores what out there in space can come to Earth or to our vicinity, along with some key influences that astronomical features have already had on life on our planet. This chapter discusses planets, asteroids, meteors, meteoroids, and meteorites—and the confusing and frequently changing terminology that astronomy employs. The next one turns to another source of Earth-bound trajectories—comets—and the even more distant reaches of the Solar System where their precursors reside.

BLURRY BOUNDARIES

My collaborators and I are chiefly theoretical particle physicists. This means that we study the properties of elementary particles—the basic ingredients of matter. Astronomers, on the other hand, focus their studies on the largest objects in the sky. They investigate what those objects are and how they coalesced from more elementary matter to evolve into what we see today. Particle physicists have been known to create fanciful terms or to usurp the names of people when christening as-yet undiscovered—and sometimes purely hypothetical—objects, such as "quarks," the "Higgs boson," or "axions." But our nomenclature seems downright methodical compared to most astronomy names, which often become a target of particle physicists' jokes. Emerging as they did from their historical context rather than an interpretation based in the science we now have at our disposal, astronomy's naming conventions and measuring units frequently come across today as arcane and confusingly unintuitive. The terms often relate to what was known or merely guessed at when something was discovered—rather than to our current understanding.

For example, you might have thought Population I would be a good way to refer to the first stars in the Universe. But Pop I already referred to one later group of stars, and Pop II was used for another. So when the group of ephemeral earliest stars was hypothesized, it was called Pop III. A comparably confusing example is the term *planetary nebula,* which is the end stage of a red giant star and has nothing to do with planets. Its perplexing name arose because the astronomer William Herschel misidentified the object he saw in his telescope when he first observed it in the late eighteenth century.

Astrophysics probably has some of the most confusing terminology because people have attempted astronomical observations for centuries, with their conclusions significantly predating any theory that would correctly explain whatever the terms refer to. Only rarely at the time of discovery did anyone grasp the more complete picture, which usually would emerge only later on. Without a better understanding, the names couldn't be rooted in a valid organizing principle.

The terminology used for planets, asteroids, and meteors is no exception. The original categories were overly broad—encompassing objects of very different types. Only after the discovery of new objects revealed the inappropriateness of the original terms could people figure this out. Even so, the original names usually survived, but with definitions that changed over time. I'm generally wary of name changes, which in business or politics are often used to divert attention from the real issues. However, most of the evolution of terminology in astromony reflects true scientific advances. The exciting phenomenon that the current proliferation of terms represents is the spectacular progress that has been made over time in our understanding of the Solar System.

PLANETS

In its initial incarnation, the term *planet* was one such liberally applied word. When the ancient Greeks first devised the term that became "planet," they were unaware of the distinctions among most celestial bodies. Scientists would have needed more sophisticated measuring tools to have recognized the differences among the identical-looking spots of light on the sky. One thing that Greek astronomers could observe was that some objects moved, so they created a separate term, *asters planetai,* or "wandering stars" for those. But the initial definition encompassed not only planets, but the Sun and the Moon as well.

Further discoveries called for further refinements in terminology. Though originally very inclusive, over time the term *planet* became increasingly restrictive. Its meaning was at first refined to refer to the five planets (aside from the Earth, which didn't qualify as a planet in a geocentric model) that are visible to the naked eye, and then later to others that were discovered with telescopes.

Planets—as we now understand the term—were created after the birth of the Sun, when dust grains collected increasingly large amounts of material that then collided, growing more or less into their current state over a time span of perhaps a few million to a few tens of million years—a very brief time interval from an astronomical perspective.

The composition and state of a planet depends on its temperature—an important influence on asteroids and comets too. As you would expect, material accreting onto planets near the Sun was a lot hotter than that collecting on planets farther out. The higher temperature kept water and methane in a gaseous state in a region that extended as far away from the Sun as four times the distance of the current location of Earth, so initially only little of that material condensed there. Moreover, the Sun sent out charged particles that swept away hydrogen and helium in the more local vicinity. So only robust materials that wouldn't melt at those temperatures, such as iron, nickel, aluminum, and silicates, could condense into the inner planets.

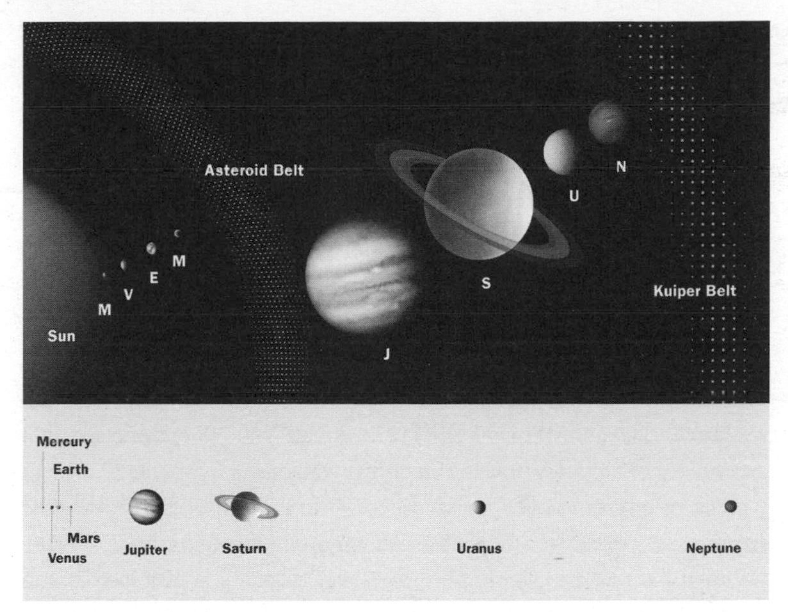

[FIGURE 9] The four rocky inner planets and the four larger outer gaseous ones with relative sizes as shown. Also indicated are the asteroid belt and the Kuiper belt. Legend below gives planet names and their relative locations in the Solar System.

Indeed, this is the material that composes the inner four terrestrial planets—Mercury, Venus, the Earth, and Mars. These elements are relatively scarce, so the inner planets needed time to grow. Collisions and mergers were essential to their achieving their current sizes, which are nonetheless small compared to those of the outer planets. (See Figure 9.)

Farther away from the Sun, between the orbits of Mars and Jupiter, lies the border beyond which volatile compounds like water and methane remain frozen as ices. Planets in this outer region grew larger more efficiently, since they are made from material that is far more abundant than those that make up the terrestrial planets. This included hydrogen, which they could accumulate in large amounts when they formed sufficiently quickly. Together, the four gas giant planets, as they are known—Jupiter, Saturn, Uranus, Neptune—

contain 99 percent of the Solar System's mass (aside from the Sun itself), with Jupiter, which is the closest to the dividing line where material could accumulate, carrying the bulk of it.

Within the past twenty years, further planet-like objects were discovered in our outer Solar System—not to mention the many others that were found in orbit about other stars. "Planet" was no longer a simple enough category, as the members of the group came to vary in size from smaller than the Moon to nearly big enough to generate the nuclear-burning characteristic of a star. Although such revisions calling for more formal definitions have happened many times before—Ceres was a planet for fifty years after its discovery before being reclassified as an asteroid—the latest debate was recent enough that many of us paid attention to the controversy as it happened.

You might remember the news stories about whether Pluto should continue to qualify as a planet. Astronomers still informally argue about the topic, and sometimes even vote to reinstate its former status. The initial debate, which was heated but somewhat arbitrary, was sparked by some relatively recent scientific discoveries. The controversy was not entirely unexpected, since people had known since the 1920s, when Pluto was first discovered, that it was weird. Its orbit is much more eccentric—elongated—than those of the other planets. And its inclination—the angle relative to the plane of the Solar System—is much bigger. Pluto was also very small compared to the other distant planets in the Solar System—the so-called gas and ice giants. It was clearly an oddity in the planet kingdom.

But it was seventy years before several similar objects were discovered in nearby orbits that showed that wacky Pluto really wasn't all that special and shouldn't necessarily have been singled out for planetary status in the first place. The argument for revising Pluto's categorization, in a nutshell, was something like the one that is used when formulating many arbitrary rules. "If we let you in, then we have to let in everyone else." It can be a lazy argument designed to avoid more subtle demarcations, and it's rarely satisfying or convincing. But objects were found that were comparable in size and orbital location to Pluto. If Pluto were to continue as a planet, so would the one similar object named Eris that had been discovered in 2005—and most likely a few others too. Eris was particularly disturbing when measurements indicated it was about 27 percent heavier than Pluto. With the threat of more such discoveries looming, someone (or some organization) would have to decide what the mass cutoff on the lower end would be for an object to achieve planetary status. But demote Pluto and the problem is solved. And that's what the International Astronomical Union (IAU) decided to do in their 2006 General Assembly in Prague. They followed the playbook for what people do in situations like this. They changed the rules of entry.

So now a *planet* is classified as an object that is round due to its own gravity and has "cleared its neighborhood" of smaller objects that would otherwise orbit the Sun in its vicinity. That means that objects such as Pluto and Eris, which are part of belts of nearby objects that orbit nearby independently, no longer rank as planets. Objects like Mercury and Jupiter, on the other hand, are roughly spherical and isolated in their orbits. Though very different from each other, they both therefore qualify.

This means that although many of us were born into a world with nine Solar System planets, we now live in a world with only eight. You might find this frustrating, but probably not to the extent experienced by those attending college in the United States in 1984, who were demoted from legal drinking age to underage by a change of law on July 17 of that year. Pluto was similarly demoted in 2006 when the IAU changed the planetary entry rules.

Interestingly, the initial estimate of the relative sizes of Eris and Pluto turned out to be misleading. Though Eris was thought to be bigger than Pluto, the margin of error was so large that astronomers had to wait for more detailed views to verify this claim. The New Horizons spacecraft, which did an up-close flyby of Pluto in July 2015 that yielded spectacular images and more detailed information, showed the size (if not the mass) of Pluto was in fact bigger. If this ambiguity had been clear from the very beginning, Pluto might still be among the ranks of the elite.

As a consolation prize, at the same meeting where "planet" was (re)defined, the IAU invented the term *dwarf planet* for objects like Pluto that (metaphorically) fell between the cracks of asteroids and planets. Pluto became the first member and the exemplar of this newly created club. The specific name *dwarf planet* was and is a subject of controversy, since—unlike dwarf stars, which are in fact stars—dwarf planets are not actually planets. The name of course arose because originally the distinction wasn't clear. The other proposed names are perhaps even more ridiculous, though, such as "planetoid" or "subplanet."

Like planets, dwarf planets have to orbit the Sun and not rotate like a moon around another planet. They are unlike asteroids in that they are not just arbitrarily shaped rocks. According to the definition, dwarf planets, which are larger than asteroids, must be massive enough to become nearly spherical under their own gravity. But dwarf planets won't have isolated orbits like true planets do. Many other objects will orbit nearby. It is only this lack of isolation—they didn't clear their vicinity—that excludes them from planetary status. An astrophysics colleague joked that planets—like senior faculty—clear nearby orbits. Dwarf planets would then be more like postdoctoral fellows, who work independently but nonetheless have offices close to the graduate students—who, like asteroids, are less well formed.

To date, dwarf planet is a rather limited category. Pluto and Ceres—the largest object in the asteroid belt but the smallest of the known dwarf planets—are the only verified dwarf planets. Ceres is furthermore the only one in the inner Solar System. The farther-out objects Haumea, Makemake, and Eris are also officially recognized, since they are sufficiently big that they are almost certainly fairly spherical, although their shape has yet to be reliably observed. Other candidates are likely to fit the bill too, such as the mysterious object Sedna, but we will only know after better measurements are completed. However, many astronomers think there are more—perhaps as many as 100 or 200 dwarf planets contained in the distant Kuiper belt, which we'll turn to soon. The Kuiper belt is probably where the objects mentioned above originated, and very likely the source of many more of a similar type yet to be discovered.

ASTEROIDS

As opposed to "planet" and "dwarf planet", the term "asteroid" remains a bit vague and colloquial since the astronomical societies never formally defined it. Even up to the middle of the nineteenth century, the words "asteroid" and "planet" were used interchangeably—and were usually taken to be synonymous. When we use the term *asteroid* today, it generally refers to an object that is bigger than a meteoroid but smaller than a planet—encompassing objects in the inner Solar System whose sizes range from tens of meters across to almost a thousand kilometers. As the appropriately named *New Yorker* writer Jonathan Blitzer described them, "Asteroids are the Solar System's most veteran castaways: rocky bodies, in orbit around the sun, left over from the formation of the Solar System. Too small to be planets, too big to be ignored, they can reveal a great deal about our primordial history."

[FIGURE 10] Images of asteroids and comets that have been visited by spacecraft as of August, 2014. Sizes range from about 100 kilometers to a fraction of a kilometer. (Composite image created by Emily Lakdawalla. Data from NASA / JPL / JHUAPL / UMD / JAXA / ESA / OSIRIS team / Russian Academy of Sciences / China National Space Agency. Processed by Emily Lakdawalla, Daniel Machacek, Ted Stryk, Gordan Ugarkovic.)

Unlike dwarf planets, asteroids are usually irregularly shaped. (See Figure 10.) The low upper limit to the observed rotation rate of asteroids leads scientists to suspect that most are not tightly bound objects, but are merely accumulations of debris, since rubble would fly apart if rotation speeds were higher. Support for this conjecture comes from spacecraft that have visited asteroids and a few observations of asteroid moons, both of which argue for low asteroid density.

Asteroids are not in short supply. There are probably billions. And they vary wildly in their composition. Most are either stony asteroids of the S-type, which are made of ordinary silicate rocks that are found mostly near Mars, or they are C-type, carbon rich, which are mostly found nearer to Jupiter. The latter receive particular attention when people consider life's origins in the Solar System since carbon is critical to life as we know it. Intriguingly, laboratory studies of meteorites demonstrate that some asteroids also contain tiny amounts of amino acids, making them even more interesting from this perspective. In the next chapter we will see that this is also true of comets, making them another important topic when considering life's origins, which I will do later on. Water too is a key component of life and some asteroids contain water, though comets generally contain more. Metallic asteroids, composed primarily of iron and nickel, exist too, but are rarer—only a few percent of the asteroid population—though

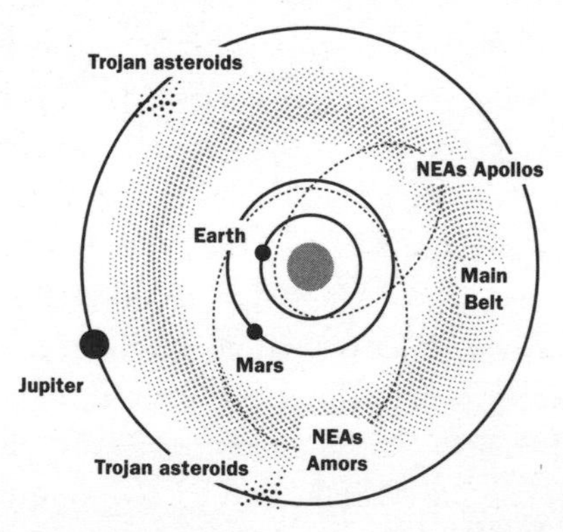

[FIGURE 11] The main asteroid belt between Mars and Jupiter, as well as Trojan asteroids, an example of an Apollo asteroid, and one of an Amor.

at least one relatively well-studied asteroid has a nickel-iron core and a basaltic crust.

As opposed to planets, asteroids are rarely alone. They orbit in specific regions in the Solar System, accompanied by many others circulating nearby. Most are located in the *asteroid belt,* which extends from Mars to the region that includes Jupiter's orbit, spanning the outer edge of the region of terrestrial, rocky planets interior to the frozen gas objects farther out. (See Figure 11.) The belt extends from about two AU to four AU—roughly 250 million kilometers from the Sun to about 600 million kilometers away. Outside the main belt, the orbits of *Trojans*, another category of asteroids, are tied to that of a larger planet or a moon—ensuring their steadiness over time.

DISTRIBUTION OF ASTEROIDS

A lot of progress has been made in the science of the asteroid belt's formation, starting in the 2000s, when astronomers began to understand planetary migration in the early Solar System. We now know that a few million years after planets began to form, charged particles ejected by the Sun eliminated most of the remaining gas and dust from the disk. Formation of the planets then ended, but not that of the Solar System. Planets moved after this time, sometimes very disruptively, scattering material out of the Solar System altogether or moving smaller objects around. One of the significant research advances in planetary science of the last few decades is an appreciation and recognition of the role this planet migration played in forming the Solar System as we now know it. The gas planets moved most significantly, affecting the development of asteroids and comets. The inner planets also migrated inward, but only a bit, and therefore probably played less of a role. It's likely that a large number of asteroids were sent into the inner Solar System when disturbed by several of the outer planets' motion outwards and Jupiter's move inward—beginning an event known as the Late Heavy Bombardment, which occurred about four billion years ago (about 500 million years after the Solar System's formation). The heavy cratering on the

Moon and Mercury from these impacts provides evidence for this event.

Astronomers think that asteroids are remnants of the protoplanetary disk that was present before planets were formed. The asteroid belt probably had far more mass initially—most of which was lost in the dynamic early days of the Solar System. Jupiter scattered away many of the objects originally in this region before they could coalesce, which probably explains the absence of planets in the region. Because it lost so much material—although the belt has many hundreds of thousands of objects bigger than a kilometer—its total mass today is a mere twenty-fifth the mass of the Moon, with the single object Ceres containing a third of that amount. When Ceres's mass is combined with that of the next three largest asteroids, half the total is accounted for, with the rest carried by the millions of smaller objects. On top of the hundreds of thousands or perhaps a million asteroids more than a kilometer wide, the belt contains even more asteroids that are smaller. Although harder to see, the numbers increase rapidly at smaller sizes, with the rule of thumb being roughly 100 times more objects that are 10 times smaller.

In the process of ejecting small planetesimals from the region of the asteroid belt, Jupiter might also have thrown some water-bearing objects to Earth. Though the origin of the Earth's water is poorly understood, these Jupiter-induced early impacts might have played some role in the Earth acquiring its abundant water supply since water could initially more readily accumulate in sufficient amounts in the outer, colder parts of the Solar System. Interestingly, "soon" (on geological scales) after the early bombardment episode ended—about 3.8 billion years ago—life began to emerge. Fortunately for its perpetuation, though smaller, less frequent impacts continue to this day, asteroids and comets don't pelt down on the planet at the dangerous rate they did early on.

Like planets, the first asteroids to be discovered were assigned various symbols. By 1855 there were a couple of dozen. Many of their names had mythological origins, but more recent discoveries assigned those fanciful names including popular culture icons—such as "James

Bond" and "Cheshire Cat"—and even the names of the discoverers' relatives. Looking at the symbols for the individual asteroids reminds me of looking at a tablet of hieroglyphics. (See Figure 12.) As my collaborator said, they resemble the name given to the "artist formerly known as Prince." The analogy is especially apt since, like Prince himself, these objects now have more readily pronounceable names.

ASTEROID	SYMBOLS		YEAR
1 Ceres		Ceres' scythe, reversed to double as the letter C	1801
2 Pallas		Athena's (Pallas') spear	1801
3 Juno		A star mounted on a scepter,for Juno, the Queen of Heaven	1804
4 Vesta		The altar and sacred fire of Vesta	1807
5 Astraea		A scale, or an inverted anchor, symbols of justice	1845
6 Hebe		Hebe's cup	1847
7 Iris		A rainbow (iris) and a star	1847
8 Flora		A flower (flora) (specifically the Rose of England)	1847
9 Metis		The eye of wisdom and a star	1848
10 Hygiea		Hygiea's serpent and a star, or the Rod of Asclepius	1849

[FIGURE 12] Names, symbols, and dates of discovery of the first ten asteroids to be found.

Scientists in early times probably approached their discoveries in a mystified state not so different from that of the ancient Egyptians. This is not to say the ancients didn't try to find order. But the Universe is a complex place in which time, dedication, and technology are essential to successfully sorting out the nature of the objects it contains. With only limited observational capacity it is difficult to tell whether an object appears dimmer or brighter or larger or smaller because of its size, composition, or location. Only time—and better measuring tools—could lead to true scientific understanding.

At the time the original "planets" were discovered, no one knew anything about asteroids. Asteroids—like planets—don't emit visible light. Planets, asteroids, and meteoroids are illuminated only by the light they reflect from the Sun. Finding asteroids is more difficult, however, since they are so much smaller and therefore dimmer and harder to see. Comets have bright trailing tails and shooting stars are relatively nearby and bright. Asteroids, on the other hand, have no readily apparent features so discovering them was (and is) a challenge.

Indeed, it took at least a couple of thousand years before people looking up realized that asteroids graced the firmament. Without extremely sensitive tools, the only way to find these dim objects was to stare for a very long time—though it also helps to know in advance where to look. The first attempts relied on this latter logic. Astronomers didn't really know the best target locations, but they used a heuristic law they thought would help guide them in their search. The Titius-Bode law that they employed seemed consistent with the location of the known planets and predicted the location of others. The discovery of Uranus in 1781 where the law said it should be seemed like a major success. Even so, there was no true theory to justify the "law," and in any event the location of Neptune doesn't conform to what it predicts.

Yet despite the arbitrariness of the suggested locations, the technique that was employed to search for planets (remember, at the time no asteroid had yet been found)—even with only eighteenth-century technology—was robust. Observers compared sky charts on different nights and searched for objects whose location had changed. Nearby planets moved noticeably, whereas distant stars appeared to be fixed. Using this method (and guided by the Titius-Bode law), Giuseppe Piazzi, founder and director of the Palermo observatory in Sicily, and also a Catholic priest, discovered an object on New Year's Day of 1801 that orbited between Mars and Jupiter, and the mathematician Carl Friedrich Gauss subsequently calculated its distance from the Earth.

We now know that the object that they had discovered, Ceres, was not a planet, but was instead the first asteroid to be found. It lies in what we now know to be the asteroid belt that is located between

Mars and Jupiter. Following several subsequent such discoveries, the astronomer Sir William Herschel suggested a separate term *asteroid* for them. The name was derived from the Greek word *asteroeidēs,* which means star shaped, since they appeared more pointlike than planets. Ceres, which we now know to be nearly spherical and about a thousand kilometers in diameter, was even more special than the other asteroids—it turned out to be the first dwarf planet to be discovered too.

Asteroids were only minimally understood until technology and the space program advanced to the point where many of these objects could be better observed. The remarkable and ongoing progress from researchers in this field has been stunning. As exciting as discovering asteroids can be, observing and exploring them is even better. The ongoing space program has designed several recent missions with this goal in mind. These more direct explorations will greatly improve on the earlier, less detailed observations that began in the 1970s, when close-up images first revealed asteroids' irregular shapes.

Other notable asteroid missions from the past include NEAR Shoemaker, the first dedicated asteroid probe, which in 2001 took photographs of, and even landed on, the asteroid Eros—the first near-Earth asteroid to be discovered. Japan's Hayabusa mission returned stony asteroid samples in 2010 and Japanese scientists recently launched the even more ambitious Hayabusa 2, which will land on an asteroid, where it will deploy three rovers to collect more samples by the end of the decade. NASA is about to launch OSIRIS-Rex, which should bring back samples of a carbonaceous asteroid.

Even more prominent in recent news has been the European Rosetta spacecraft, which flew by and gathered detailed information about the asteroids Lutetia and Steins before its more famous and more recent comet rendezvous. NASA's spacecraft Dawn is making news these days too. It has already visited Vesta and has now reached the dwarf planet Ceres.

In the future, the ambitious asteroid mining operations that are currently under consideration—though not necessarily the most obvious route to economic gain—are likely to prospect many more asteroids as well. So too might the spacecrafts designed with an eye toward asteroid deflection that are currently in the works, such as the ambitious Asteroid Redirect Mission (ARM), which NASA is developing. The current American space program's focus on asteroids—the planets' less glamorous but often more accessible counterparts—will very likely teach us quite a bit about our Solar System.

METEORS, METEOROIDS, AND METEORITES

Let's now turn from asteroids to the even smaller objects known as meteoroids. The study of meteoroids is known by the awkward-sounding term " "meteoritics"—and not "meteorology", which probably would have been the more sensible-sounding name for the study of the small stony objects in the sky. But before astronomy could claim the term, which comes from the Greek *meteoreon*—meaning "high in the sky"—and *logos*—the word for "knowledge—weather studies had already usurped it. Unfortunately for today's terminology, the ancient Greeks thought that studies of the weather fit the bill for meteorology—the study of objects in the sky.

The first standardized definition of *meteoroid,* devised only in 1961 by the International Astronomical Union, was a solid object moving in interplanetary space that was considerably smaller than an asteroid and considerably larger than an atom. Though more sensible than "meteorology" from an astronomical perspective, it is not very specific either. In 1995, two scientists suggested restricting the size to between 100 micrometers and ten meters. But when asteroids smaller than ten meters across were found, scientists at the Meteoritical Society suggested changing the size range to between 10 micrometers and one meter—about the size of the smallest asteroid ever observed. But this change never became official. I'll often use the term "meteoroid" fairly liberally for medium-sized objects in the sky, but will refer to objects that are even smaller by their more accurate names of *micrometeoroids* or *cosmic dust.*

Like asteroids, meteoroids differ dramatically in their nature, probably as a result of their wildly varying origins in the Solar System. Some are snowball-like objects with densities only a quarter that of ice, while others are dense, nickel- and iron-rich rocks, while still others have more abundant carbon.

Although colloquial use for the term "meteor" frequently includes the meteoroid or micrometeoroid that created it, the correct use of the word corresponds to the term's Greek root, which means "suspended in the air" and refers only to what we see in the sky. A *meteor* is the visible streak of light that is produced when a meteoroid or a micrometeoroid enters the Earth's atmosphere. Despite this definition, most people—even reporters—incorrectly talk about meteors falling to Earth, as does the name of the universally declaimed 1979 film *Meteor,* which in fairness has its entertaining moments.

Amusingly, like "meteorology," the term "meteor" had an early definition connected to weather—originally encompassing any atmospheric phenomenon, such as hail or a typhoon. Winds were called "aerial meteors," rain, snow, and hail were called "aqueous meteors," light phenomena such as rainbows and aurora were called "luminous meteors," and lightning and what we now call meteors were called "igneous meteors." These terms are relics from a time when no one knew how high anything was or that weather features had dramatically different origins from astronomical ones. "Meteorology" as a term is perhaps not entirely misguided since weather is indeed related to our position in the Solar System—but of course very differently than originally conceived. Happily, despite "meteorologists'" earlier misunderstandings, the term "meteor" is no longer used in this way.

We readily see meteors because the objects that created them heat up on entry and emit glowing material that we see as light—which appears in an arc because of the meteoroid's fast speed. Though many meteors happen randomly, meteor showers are more regular occurrences that arise from the Earth passing through comet debris. Of course meteors are more readily observed at night when the Sun's light doesn't obscure them. There is no philosophical tree-falling-in-the-forest quandary here. The existence of meteors doesn't depend on observers actually looking. The streaks of light just have to be visible in principle.

Most meteors arise from dust or pebble-sized objects. Millions of them enter our atmosphere daily. Since most meteoroids fall apart above 50 kilometers in altitude, meteors typically occur between about 75 and 100 kilometers above sea level in what is called the mesosphere. Though the precise speed depends on an object's specific properties and its alignment of the velocity with respect to the Earth, the speeds of the objects creating the meteors are generally on the order of tens of kilometers per second. The trajectory of a meteor helps identify where the meteoroid that created it came from, whereas the spectrum of visible light a meteor emits and its influence on radio signals both help scientists determine a meteoroid's composition.

The meteoroids that make it through the atmosphere and hit the Earth can lead to *meteorites*. Meteorites are the rocks left on Earth after an extraterrestrial object has hit, decomposed, melted, and partially vaporized. Meteorites are yet another tangible reminder that the Earth is intrinsically part of a cosmic environment. You might be lucky and find a meteorite in the vicinity of a meteoroid impact, but you are more likely to see them in laboratories, museums, or the houses of sufficiently obsessive, lucky, or wealthy people. The Vatican Observatory Museum has quite a nice collection, as does the Smithsonian's American Museum of Natural History, which contains the largest one. A three-star general told me that the Department of Defense has a nice assortment too. Their collection is connected to missile defense, but unfortunately their data on meteoroid impacts remain classified. The study of those meteorites that scientists can access has taught us a great deal about the Solar System and its origins.

Meteorites can also arise from comets—the objects of the outer Solar System that feature in the following chapter. Objects that orbit in the inner Solar System are different enough from the ones in the outer Solar System that only the ones inside Jupiter's orbit are referred to as asteroids or *minor planets,* which—unlike asteroid—is the somewhat dismissive-sounding but official term. The distinction between comets and asteroids might seem obvious—the most striking difference being the comet's prominent tail—but the delineation is actually more nuanced. Comets generally have more elongated orbits, but some asteroids have similarly eccentric orbits too—probably because they were comets originally. Moreover, the water-containing asteroids are not necessarily a distinct population from the comet-forming populations of the outer Solar System. This varied composition of asteroids also indicates that the population considered asteroids and the one considered comets have some overlap.

The distinction is sufficiently blurry that in 2006, the International Astronomical Union devised the term small Solar System body to encompass them both—but not dwarf planets. Dwarf planets could have been included as small Solar System bodies too, but since they are larger and more spherical—indicative of the stronger gravity that makes them more likely to be solid objects—the IAU chose to omit them from the category and distinguish them with their own name. In general, the IAU prefers the term "small Solar System body" to "minor planet" precisely because objects in the asteroid belt can sometimes have the characteristics of comet nuclei. A single term that encompasses them both—though less informative—prevents mistakes. Even so, asteroids are generally rockier and comets usually contain more volatile substances, so most astronomers persist with the distinction.

This awkward terminology leaves me in a quandary, however, when, in the rest of this book, I refer to big objects that hit the Earth. Small objects that burn up in the sky are meteoroids or micrometeoroids. Bigger objects, which originate as either asteroids or comets, occasionally make it to the Earth or its atmosphere too. But we know which one only if we observe the trajectory and hence its origin. We need a term that can refer to either. The cumbersome term "small Solar System body" technically fits the bill but is rarely used for objects flying in from the sky—especially those that come close to or hit the Earth. Check the headlines and you will find "meteor," "meteoroid," or even "meteorite" frequently employed—even though technically they are all wrong if the object is bigger than a meter. Since there seems to be no common colloquial term that is sufficiently specific (though "impactor" and "bolide" are sometimes used), I will call the objects "meteoroids"—the least egregious of these offenses—throughout this book when referring to an extraterrestrial object that enters the atmosphere or that hits the Earth. This is a slight abuse of terminology in that the term usually refers only to a smaller object. But in context it should be clear what I mean.

7

THE SHORT, GLORIOUS LIVES OF COMETS

If you ever have the opportunity to travel to the Italian city of Padua, be sure to visit the Scrovegni Chapel. This well-preserved gem from the early fourteenth century houses a magnificent fresco cycle by the early Renaissance artist Giotto. My favorite image—and the one that is treasured by all my physics colleagues there—is *The Adoration of the Magi* (see Figure 13), which shows a comet conspicuously passing over the classic nativity scene. Perhaps, as the art historian Roberta Olson has suggested, the comet replaces the Star of David—the more familiar compositional element—with a bright dramatic object that people had witnessed overhead only a few years before the painting was completed. Independent of the allegorical intention, the flash of light over the crèche is unmistakably a comet—very likely Halley's comet, which anyone in that part of the world would have seen. The enormous comet tail that extended over a significant part of the sky in September and October 1301 would have been a spectacular sight—especially in an era predating electric lights.

I like to reflect on early-fourteenth-century Italians gazing upward and appreciating the same astrophysical wonders that we marvel at today. Evidence from ancient Greek and Chinese civilizations indicates that people had been observing and appreciating comets at least two thousand years before then as well. Aristotle even tried to

[FIGURE 13] *Adoration of the Magi* by Giotto, with a comet over the traditional crèche.

understand the nature of comets—interpreting them as phenomena in the upper atmosphere where dry, hot material would start to burn.

We've come a long way since the time of the ancient Greeks. Scientists' more recent insights based on mathematics and far better observations have taught us that comets are cold, nothing is burning, and the volatile material they contain is readily converted to gaseous vapor or water once they get close enough to the Sun.

Now that we've explored the nature of asteroids, which come from relatively nearby in the Solar System, let's move on to comets, which arise from distant domains known as the scattered disk—which overlaps the Kuiper belt—and the Oort cloud—which lies in the outer reaches of our Solar System. Comets occur in other stellar systems as well. But our focus here will be on those we know best—the ones that originate in our own domain.

THE NATURE OF COMETS

Though we now know that comets arise from distant regions and only rarely follow paths that bring them close to Earth, Tycho Brahe's conclusion in the sixteenth century that comets reside outside the Earth's atmosphere was an important early milestone in scientific understanding. Tycho measured the parallax of the Great Comet of 1577 by integrating sightings from observers in different places—thereby determining that comets were at least four times as far away as the Moon. That's an understatement for sure but it was a big leap forward at the time.

Isaac Newton made another important deduction when he realized that comets move in oblique orbits. He demonstrated by using his gravitational inverse square law, which says the strength of gravity is four times smaller for an object twice as distant—that objects in the sky have to follow elliptical, parabolic, or hyperbolic orbits. When Newton fit the path of the Great Comet of 1680 to a parabola, he literally connected the dots—showing that the objects people had sighted and believed were different in fact fell along a single trajectory—the path of a single thing. Though the comet's trajectory is actually an elongated ellipse, the path was close enough to parabolic in shape that Newton's deduction of a single, moving object remained correct.

Early on, the first comets to be discovered were named after the year they appeared. In the early twentieth century, the naming convention changed, and they became the namesake of the person who predicted the orbit, such as the German astronomer Johann Franz Encke and the German-Austrian military office and amateur astronomer Wilhelm von Biela—both of whom have comets carrying their name.

Though identified well before the twentieth century, Halley's comet too was named after the man who understood its trajectory sufficiently well to predict its recurrence. In 1705, using Newton's laws and taking into account perturbations by Jupiter and Saturn, Newton's friend and publisher Edmond Halley predicted that a comet that had already appeared in 1378, 1456, 1531, 1607, and 1682 would reappear in 1758–59. Halley was the first to suggest a comet's periodic motion and he got it right. Three French mathematicians did even more precise calculations and predicted the 1759 date to within a month. We can do similar calculations today to determine that not before 2061 will someone on Earth witness Halley's comet again.

Later in the twentieth century, the convention changed again, converting to naming comets after the people who discovered them. And once comet discovery became a collective effort based on more advanced observational tools, comets were named for the instrument that found them. The current list contains about 5,000 comets, but a realistic estimate of their total number is at least a thousand times greater and could be far bigger still—perhaps as high as a trillion.

Understanding the nature and composition of comets requires a little knowledge about the states of matter. The familiar phases of matter are solid, liquid, and gas—which for water amounts to ice, water, and steam. Atoms are arranged differently in each of the phases, with solid ice the most structured and gaseous vapor the most random. When a phase transition converts liquid into gas—as happens when water boils, for example—or solid into liquid—as occurs when ice cubes melt—the material stays the same since all the same atoms and molecules are present. But the material's nature becomes very different. The form that matter takes depends on its temperature and its composition—which determines the boiling and melting points for any particular stuff.

I was amused to hear that someone recently took advantage of the different phases to try to bring a bottle of water past airport security. He froze it and argued that the solid ice in his bottle didn't defy any anti-liquid ban. Unfortunately, the Transportation Security Administration agent was not convinced. Had the representative been educated in physics, he might have compellingly argued that only substances that are solid under standard temperature and pressure are allowed. However, I'm pretty sure this isn't what was said. (Note that temperature and physics both play a role because melting and boiling points are different at different pressures, as anyone who has tried to make pasta in Aspen, Colorado, at eight thousand feet above sea level, would know.)

The melting and boiling points are critical to any structure since they determine the phase that its material will take. Some elements, such as hydrogen and helium, have extremely low boiling and melting points. Helium becomes liquid, for example, only at four degrees above absolute zero. Planetary scientists call those elements with melting point below 100 degrees kelvin gases—independent of the actual phase the matter is in. Those with low melting point, but not so low as gases, are referred to—again by planetary scientists—as ices, though whether or not the material is actually an ice also depends on the actual temperature. That is why Jupiter and Saturn are called gas giants, while Uranus and Neptune are sometimes called ice giants. In both cases, the interior is actually a hot, dense fluid.

Gases (in the sense used by planetary scientists) are a subset of *volatiles,* which are elements and compounds with low boiling points—such as nitrogen, hydrogen, carbon dioxide, ammonia, methane, sulfur dioxide, and water—that might be present in a planet or atmosphere. A material with low melting point will more readily turn into a gas. You might have seen ice cream made from cold liquid nitrogen. (This is a staple of molecular cuisine in modern restaurants—as well as a standard science fair demonstration. It also featured in one of the food trucks outside Harvard's Science Center, which, fortunately for my health, usually made ice cream flavors I don't like.) If you have seen any such example, you will have noticed how readily the nitrogen atoms escape as gas at room temperature, making the apparatus look very dramatic (and a little like a caricature of a lab experiment).

The Earth's Moon is low in volatiles, since it consists mostly of silicates, with little in the way of hydrogen, nitrogen, or carbon. Comets, on the other hand, contain abundant volatiles, which is what gives rise to their dramatic tails. Comets originate well beyond Jupiter in the outer regions of the Solar System, where water and methane remain cold and frozen. In these very cold regions far from the Sun, ice doesn't turn to gas. Ice remains ice. Only when comets pass into the inner Solar System where they are closer to the heat of the Sun do the volatile materials in the comet vaporize so that they stream out, along with some dust, creating an atmosphere around the nucleus called the *coma*. The coma can be much bigger than the nucleus—thousands or even millions of kilometers across, sometimes even growing to the size of the Sun. Bigger dust particles remain in the coma whereas lighter ones are pushed into the tail by the Sun's radiation and charged particle emissions. A comet consists of the coma, the nucleus it surrounds, and the tail that streams away.

Meteor showers, which arise out of the solid debris that comets leave in their wake, are spectacular evidence of comets. They occur after a comet has crossed the Earth's orbit so that some of the discarded material lies along the Earth's path. The Earth then passes through the debris on a regular basis, creating the wonderful periodic meteor showers that are so remarkable to see. Comet Swift-Tuttle's debris is the origin of the Perseid meteor shower, which occurs in early August, and that I've unwittingly encountered in the clear skies of Aspen, where a physics center hosts summer workshops. Another example is the Orion meteor shower, which occurs in October and arose from the scattered fragments of Halley's comet.

Comets are among the most spectacular objects we can witness overhead with the naked eye. Most are very faint, but ones such as Halley's, which are visible without a telescope, pass by perhaps a few times a decade. Comets orbit the Sun with bright ion tails and separate dust tails that generally point in different directions. These trails of brightly shining dust and gas are the origin of their name, which comes from a Greek word that translates as "wearing long hair." Whereas the dust tail generally follows the comet's path, the ion tail points away from the Sun. The ion tail forms when solar ultraviolet radiation hits the coma, tearing off electrons from some of the atoms there. The ionized particles create a magnetic field in what is known as a *magnetosphere*.

Something called the *solar wind* plays an important role in this manifestation of comets. Everyone is familiar with the Sun's radiation, which gives rise to the photons we experience as heat and light on Earth. Less well known are the charged particles—electrons and protons—that the Sun emits and that make up the solar wind. When in the 1950s the German scientist Ludwig Biermann (and independently another German, Paul Ahnert) made the extraordinary observation that the bright ion tail of a comet always points away from the Sun, Biermann proposed that the Sun emits particles that "push" on the comet tail, making it point this way. In a metaphorical sense, the "solar wind" "blew" the ion tail there. Understanding this process taught scientists about both comets and the Sun—and enlightened me as to the origin of the mysterious name.

Comet tails can extend up to tens of millions of kilometers. The sizes of comet nuclei are of course much smaller but still pretty big compared to a typical asteroid. The nuclei don't have sufficient gravity to round out their structure, so they have irregular shapes, varying in size from a few hundred meters to tens of kilometers across. That might be observational bias, since bigger ones are easier to spot, but searches employing sufficiently sensitive instruments to find smaller objects have so far come up empty-handed.

In terms of visibility, it's a good thing that comets have comas and tails. Comet nuclei are very nonreflective, which makes them extraordinarily difficult to see, since the most common way to view nonburning objects (such as you and me) is through reflected light. To mention one well-known example, the nucleus of Halley's comet reflects only about 1/25th of the light that hits it. This is comparable to the reflectivity of asphalt or charcoal, which we know to be very dark. Other comet nuclei reflect even less. In fact comet surfaces seem to be the darkest ones in the Solar System.* Whereas the more volatile, lighter compounds are removed by the heat of the Sun, the darker, larger organic compounds remain. The dark materials absorb light, heating the ices that send out gases to become the tails. The similarity in reflectivity between charcoal and comets is not coincidental—remember that tar is also made from large organic molecules, namely those in petroleum. Now imagine asphalt in the

---

* Note that *dark* here has the usual meaning of absorbing light. This is not "dark matter"

sky billions of kilometers away. Without a lot of effort devoted to seeking it out, such a dark object would indeed be lost in obscurity.

When comets are in the outer Solar System, they are dark and frozen with just the faintest of optical emissions. The only way to observe comets before they approach the Sun is through the infrared light they emit. Not until they get to the inner Solar System do the coma and tail form to make comets more readily visible. Dust then reflects sunlight and the ions make the gases glow, leading to light we can more readily observe. Even so, most comets are visible only with a telescope.

The exact chemical makeup of comets is even harder to observe than the objects themselves. Meteorites found on Earth give us some clues by conveying some of their actual material to our home turf. Scientists have also noted the various colors of comets and observed some of their spectral lines. Using these and other sparse clues, scientists have concluded that the nucleus consists of water ice, dust, pebble-like rocks, and frozen gases including carbon dioxide, carbon monoxide, methane, and ammonia. Nuclei surfaces appear to be rocky, with ice buried a little below the surface.

Given the limited astronomical observations of his time, Isaac Newton gave a remarkably accurate interpretation of comets in the seventeenth century. He mistakenly thought they were compact, durable solid bodies, but he recognized that the tails were thin streams of vapor that had been heated by the Sun. In terms of understanding the composition of comets, the philosopher Immanuel Kant in 1755 did an even better job. He surmised that comets were composed of volatile material that vaporizes to create the tail. In the 1950s, Fred Whipple of the Harvard astronomy department, and a discoverer of six comets himself, famously recognized the preponderance of ice in comets, with dust and rock only secondary, giving rise to the "dirty snowball" model you might have heard about. In fact the composition isn't entirely settled and some are dirtier than others, but better observations are still advancing our knowledge.

One further fascinating compositional feature of a coment is that it contains organic compounds, such as methanol, hydrogen cyanide, formaldehyde, ethanol, and ethane, as well as long-chain hydrocarbons and amino acids, the precursors of life. Meteorites on Earth have even been found to contain components of DNA and RNA that presumably came from either asteroids or comets. Objects that carry water and amino acids and hit the Earth on a regular basis are certainly worthy of our attention.

Comets' fascinating structure and possible relevance to life have made them an obvious target for a number of space missions. The first spacecraft studying comets flew by their tails and the surfaces of their nuclei to collect and analyze dust particles and perhaps take photographs, but without the proximity or resolution to provide significant detail. In 1985, the International Cometary Explorer, a redirected mission of NASA with some European support, was the first spacecraft to approach the tail of a comet, but from only 3,000 kilometers away. The Halley Armada, consisting of the two Russian-launched Vega missions, Japan's Suisei mission, and the European Giotto spacecraft, followed soon afterward to try to better study the comet's nucleus and coma. But the Giotto robotic mission—named for the painter of the comet-illustrating *Adoration of the Magi,* mentioned earlier—surpassed them all. This spacecraft approached the nucleus of Halley's Comet within 600 kilometers.

More recent missions, which have tried to directly explore comets and their composition, have done even better. The Stardust spacecraft collected and analyzed dust particles from the coma of the comet Wild 2 in early 2004 and brought this material to Earth for study in 2006. The comet's material did not consist primarily of interstellar medium material as expected for an object formed in the distant Oort cloud, but was instead mostly stuff heated from within the Solar System. Scientists showed that the comet contained minerals of iron and copper sulfide, which couldn't have formed without liquid water—implying that the comet must have initially been warmer and therefore must have formed closer to the Sun. The results furthermore demonstrated that the composition of comets and asteroids is not always as different as scientists had expected.

And Deep Impact is not just an ambitious (if somewhat confused) movie—it is also the name of a space probe that in 2005 sent an impactor into comet Tempel 1. The spacecraft was designed to study the comet interior and photograph the impact crater—though the dust cloud that the impact created somewhat obscured the images. The discovery of crystalline material, which requires much more extreme temperatures to form than comets currently experience, indicated either that material entered the comet from the inner Solar System or that the comet initially formed in a region distant from its current location.

More recent probes of comets are more exciting still. In a remarkable development, the European Space Agency in 2004 launched a spacecraft named Rosetta to orbit the comet 67P/Churyumov-Gerasimenko and subsequently land an even more direct probe named Philae onto its surface to study the nucleus composition and inner regions up close. One of the major news stories of November 2014 featured Philae, which landed—but not as smoothly as planned—with bounces sending it to a less stable location. The event, which was quite literally a cliffhanger, has accomplished a fair fraction of its scientific goals. Though the drill mission didn't succeed, Philae—even in the wrong place and without the intended attachment mechanisms in place—has studied a comet's shape and atmosphere in more detail than ever before.

Rosetta now orbits the comet and will continue to do so as it enters the inner Solar System. This entire mission is already a rather spectacular accomplishment—perhaps more impressive in light of its launch happening within a century of the Wright brothers' launch of their first plane.

SHORT AND LONG PERIOD COMETS

Yet even with all the progress, many intriguing questions about comets remain. In addition to better determining what they are made of, astronomers would like to deepen their understanding of both the orbit of comets and the manner in which comets are formed. In fact we don't necessarily expect a single unified explanation since there is evidence for distinct classes of comets, distinguished as *short-period* and *long-period* type according to the time it takes for them to complete their journey around the Sun. The demarcation between short and long periods is taken to be 200 years, but periods vary overall from a few years to several million.

Comets originate beyond Neptune and the reservoir of these trans-Neptunian objects resides in distinct orbital bands located at different distances from the Sun. The inner regions, which give rise to short period comets, are called the *Kuiper belt* and the *scattered disk,* while much farther out is the hypothesized *Oort cloud,* which produces long period comets and to which I'll soon (figuratively) return. An additional region that astrophysicists propose but that we will not focus on here lies between the scattered disk and the Oort cloud and is given the patronizing-sounding name *detached objects.*

To a large extent, the categorization of the inner and outer regions from which comets originated overlaps with their orbital period. The comets we see most often are short-period comets, such as Halley's comet, which recurs at manageable intervals and which generations of humans have observed. Short-period comets come from closer-in regions and long-period comets mostly arrive from farther out. We occasionally see long-period comets too, but only if and when they enter the inner Solar System, which might be caused by perturbations to the distant Oort cloud. The Sun's gravity binds the comets there only weakly so that even small disturbances can send objects out of their orbits, plunging inward toward the Sun. Even short-period comets, such as Halley's comet, might have first been kicked out of a more distant long-period orbit into a shorter-period one in the inner Solar System.

Short-period breaks down into two subcategories: Halley family comets, with periods greater than 20 years; and Jupiter family comets, with smaller periods. There are likely some asteroids or dormant/extinct comets on short-period orbits too, but probably extremely few asteroids have an orbital period greater than 20 years. Longer-period comets are more *eccentric*, meaning they have more elongated orbits than shorter-period ones. This makes sense, since comets are visible to us only near the Sun. Whereas short-period comets circle closer in to the Sun's vicinity, a comet that is both observable and has a long period should have an orbit that dips inward close to the Sun but then extends far outward to create the long path that requires a long time to travel around. The orbits of long-period comets also seem to lie closer to the *ecliptic plane* in which the planets travel and furthermore orbit in the same general direction.

The fate of any of these objects once they have entered the inner Solar System depends on possible further perturbations. Jupiter is the biggest known relatively local perturber since its mass is more than double the total mass of all the other planets combined. New comets in the inner Solar System might enter a new orbit or they might appear only once before getting kicked out of the Solar System or colliding with a planet, as Shoemaker-Levy famously and gloriously did when it crashed into Jupiter not so long ago—in 1994.

THE KUIPER BELT AND THE SCATTERED DISK

Let's now consider the domains that contain the icy minor Solar System bodies that will turn into comets if perturbed to enter the inner Solar System. Our first subject will be the Kuiper belt. (See Figure 14.) Though not itself the reservoir of short-period comets, it is an important guidepost to the scattered disk, which is.

To me, one of the most interesting aspects of the Kuiper belt, which was predicted in the 1940s and 1950s, is the recentness of its discovery. No further back than 1992, astronomers determined that our understanding of the Solar System, which many of us learned in grade school and thought to be on very solid footing, had to be revised to account for the discovery of the Kuiper belt and several other advances I'll soon discuss. Even if you've never heard of the Kuiper belt, you might be familiar with a few objects that reside in or originated there. These include three dwarf planets—among them the former planet known as Pluto. Though now located far from the Kuiper belt, Neptune's moon Triton and Saturn's moon Phoebe also have sizes and compositions indicating they too began their existence in this locale before planetary transits pulled them away.

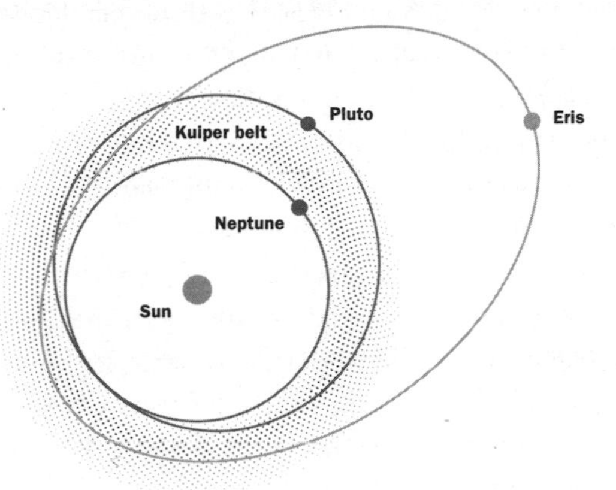

[FIGURE 14] The Kuiper belt located beyond Neptune includes Pluto as its largest object. The scattered disk, slightly outside the Kuiper belt, contains the even more massive object, Eris.

An AU, or astronomical unit, is about 150 million kilometers—the approximate distance between the Earth and the Sun. The Kuiper belt lies in a region greater than 30 times as far from the Sun—between about 30 and 55 AU away. It contains a large number of minor planets, most of which are situated in the *classical Kuiper belt,* which lies between about 42 and 48 AU from the Sun. This region extends vertically about 10 degrees outside the ecliptic plane, though the mean position has only a couple of degrees inclination. Its thickness makes it more doughnut-shaped than belt-shaped. Even so, its slightly misleading name survives.

The name is a bit unfair for another reason too. The number and variety of prior speculations on the nature of the Kuiper belt makes it unclear who precisely deserves credit for the proposal. Soon after its discovery in the 1920s, many astronomers suspected that Pluto was not alone. As early as 1930, scientists introduced various hypotheses about additional trans-Neptunian objects, but the astronomer Kenneth Edgeworth probably deserves the most credit. In 1943, he argued that the material in the early Solar System in the region beyond Neptune was too diffuse to create planets and would instead become a group of smaller bodies. He continued to speculate that on occasion one of these objects would enter the inner Solar System to become a comet.

Scientists now favor a scenario very similar to Edgeworth's, whereby the early solar disk condensed into objects smaller than planets, sometimes known as *planetesimals*. Gerard Kuiper, for whom the Kuiper belt is named, made his hypothesis later—in 1951—and furthermore didn't quite get it right, thinking it to be a transient structure that would have disappeared by today, since he thought Pluto was bigger than it actually is and would therefore have cleared the local region as other planets have done. Because Pluto is considerably smaller than Kuiper anticipated, this didn't happen, so the Kuiper belt, with its many objects in the same general region as Pluto's orbit, has indeed survived.

Edgeworth is sometimes credited along with Kuiper for his speculation through the name *Edgeworth-Kuiper belt*. But, as almost always happens with elongated names—in America at least—the shorter form is more commonly used. As with the "Sveriges Riksbank Prize in Economic Sciences in Memory of Alfred Nobel," which was created as an afterthought to the true Nobel Prizes, but is usually referred to as the "Nobel Prize in Economics" rather than by its full and more cumbersome name, you rarely hear the longer version of the astronomical term that appropriately recognizes Edgeworth's contribution.

Following Edgeworth's and Kuiper's suggestions, scientists realized that comets themselves clue us in to the Kuiper belt's existence. There were too many short-period comets discovered in the 1970s to be accounted for by the Oort cloud—the much more distant reservoir of comets that we will turn to soon. Short-period comets arose near the plane of the Solar System, as opposed to Oort cloud comets, which were far more spherically distributed around the Sun. Based on this evaluation, the Uruguayan astronomer Julio Fernandez thought the comets could be explained instead by a belt located in the region where we now know the Kuiper belt to be.

As always, despite all these speculations, discovery required the necessary observational sensitivity. Because finding small, distant, nonluminous objects is not easy, the first objects aside from Pluto in the Kuiper belt were discovered only in 1992 and early 1993. Jane Luu and David Jewitt, who started their search when Jewitt was a professor at MIT and Luu was a student, conducted their observations at the Kitt Peak National Observatory in Arizona and the Cerro Tololo Inter-American Observatory in Chile. They continued after Jewitt had moved to the University of Hawaii, where they could use the university's 2.24-meter telescope on the summit of the now-dormant volcano Mauna Kea—a beautiful viewing site with wonderfully clear skies (well worth visiting if you are on the Big Island). After five years of searching, they discovered two Kuiper belt objects—the first in the summer of 1992 and the other early in the following year. Since then, many other such objects have been discovered, though they almost certainly represent only a tiny fraction of what exists. We now know the belt contains more than a thousand residents, which are known as Kuiper belt objects, or KBOs, though calculations argue that as many as a hundred thousand with diameters greater than 100 kilometers could be out there.

It is worth noting that despite Pluto's loss of planetary status, it is still special, which is why it was discovered before any other object in the Kuiper belt. Based on what we now know about the masses of objects in its vicinity, Pluto is larger than would be expected. This one object seems to carry a few percent of the total mass of the Kuiper belt and is very likely the largest such object there. In fact, the Kuiper belt's low net mass is an interesting clue to its origin. Although estimates range from about 4 to 10 percent of the Earth's mass, Solar System formation models would assign the Kuiper belt more like 30 times the mass of the Earth. If its mass had always been so low, no object greater than 100 kilometers in diameter would have joined the belt—which is contradicted by Pluto's existence. This tells us that a large fraction—more than 99 percent of the predicted mass—isn't there. Either KBOs formed elsewhere—closer to the Sun—or something dispersed most of the mass.

The many other objects that have a similar orbit to Pluto are called *plutinos* and are located a little less than 40 AUs from the

Sun, though their highly eccentric orbits mean their distances vary. Plutinos are *resonant Kuiper-belt objects,* which are those that travel in orbits that have a fixed ratio with respect to Neptune. Plutinos, for example, orbit the Sun twice during the time Neptune makes three trips around. The fixed ratio prevents the objects from getting too close to Neptune, and so they escape its stronger gravitational field that would otherwise kick them out of the region. Amusingly, the International Astronomical Union (IAU) requires that plutinos, like Pluto, have also to be named after underworld deities. We know of at least a thousand such objects, though, given the limited surveys to date, scientists suspect—as with the other categories I've discussed—the existence of far more.

However, the dominant Kuiper belt population does not consist of plutinos, but of objects that lie in the classical Kuiper belt. Surveys have found many such objects, and the Pan-STARRS survey project, which is now searching full-time for anything in the Solar System that visibly moves, will very likely find many more. The objects in the classical Kuiper belt have stable orbits that aren't disturbed by Neptune—even without any resonant orbit that keeps it a fixed distance away. A good fraction of these redder classical objects have very circular orbits. A second population has more eccentric and more inclined orbits—up to a maximum of about 30 degrees—but typically far less. This leaves some relatively unpopulated unstable regions within the Kuiper belt, containing only those objects that have arrived fairly recently.

Objects that used to be in the Kuiper belt are likely precursors or at least related to many of the comets we observe, so it shouldn't be surprising that their composition is essentially that of comets. They are mostly made of ices of materials such as methane, ammonia, and water. The presence of ice rather than gas is due to the location of the belt and its consequent low temperature of about 50 degrees kelvin—more than 200 degrees colder than the freezing point of water. After scientists finish analyzing data from the New Horizons spacecraft, which will have gathered a lot of information about Pluto and the Kuiper belt, we should learn a good deal more.

The orbits in the belt are stable, however, so comets don't originate precisely there. Kuiper belt permanent residents won't make their way to the Sun. Instead, short-period comets arise from the *scattered disk*—a relatively empty region containing icy minor planets that overlaps with the Kuiper belt but extends much farther from the Sun—to 100 AU or more. The scattered disk contains objects whose orbits can be destabilized by Neptune. The greater eccentricity, range of locations, and degree of inclinations—up to about 30 degrees—distinguish the scattered disk population from that of Kuiper belt objects, as does their instability. Scattered disk objects have medium to high eccentricity, meaning they have stretched-out, rather than circular, orbits. Their eccentricity is so high that even objects whose maximum range is far away from Neptune get close enough during their orbits to be subject to Neptune's gravitational field. This is why the influence of Neptune can sometimes send scattered disk objects into the inner Solar System, where they get warmed up to release gas and dust and thereby become recognizable as comets.

Eris, the one known minor planet comparable in size to Pluto, lies outside the Kuiper belt in the scattered disk and so is the unpoetically named (48639) 1995 TL_8, which was discovered in 1995 but only later classified as a scattered disk object. To find it, astronomers on Mauna Kea used charge-coupled devices—an advanced version of the technology used in digital cameras—along with better computer processing. These made more distant objects possible to observe and contributed to Eris's quite recent 2005 discovery. Astronomers found three more scattered disk objects a few years later. Hundreds more have been discovered since then. Their total number is probably comparable to that of Kuiper belt objects, but being farther away they are more challenging to observe.

Kuiper belt objects and scattered disk objects contain similar compounds. Like other trans-Neptunian objects, the objects in the scattered disk have low density and are composed primarily of frozen volatiles such as water and methane. Many think that Kuiper belt objects and scattered disk objects started off in the same region, but gravitational interactions—primarily with Neptune—sent some into stable orbits in the Kuiper belt, and others inward into a region containing objects called Centaurs which lies between the orbits of Jupiter and Neptune. Gravitational interactions sent the remaining objects into the unstable orbits of the scattered disk.

The gravitational influence of the outer planets is almost certainly responsible for much of the structure of the Kuiper belt and the scattered disk. It appears that at some point Jupiter drifted inward, toward the center of the Solar System, while Saturn, Uranus, and Neptune moved outward. Jupiter and Saturn used each other to stabilize their orbits—Jupiter orbits the Sun exactly twice as fast as Saturn. But these planets destabilized Uranus and Neptune—putting them into different orbits, with Neptune becoming more eccentric and orbiting farther out. En route to its final destination, Neptune likely scattered many planetesimals into more eccentric orbits and many others into more inner orbits where they would rescatter or get ejected by Jupiter's influence. This would have left less than one percent of the Kuiper belt intact, while the majority would have scattered away.

A competing proposal is that the Kuiper belt formed first and the scattered objects came from the Kuiper belt. In this proposal—similar in many respects to the one above—Neptune and the outer planets scattered some objects into eccentric and inclined orbits, either toward the inner solar region or outward into farther reaches of the Solar System. Some of the objects scattered from the Kuiper belt outward would then have become scattered disk objects. Others might have become Centaurs. This would solve the mystery of how the Centaurs, which have unstable orbits and can stay in their domain only a few million years, can exist to this day—the Kuiper belt might replenish them. Comets, too, have only finite (but glorious) lifetimes. The Sun's heat gradually erodes them by sublimating their volatile surfaces. Without a continuous source of new objects, comets would no longer be around.

THE OORT CLOUD

The scattered disk is the reservoir of short-period comets. The Oort cloud, an enormous spherically distributed "cloud" of icy planetesimals containing perhaps a trillion minor planets, is the hypothesized source of long-period ones. (See Figure 15.) The Oort cloud is named after the Dutch astronomer Jan Hendrik Oort, who has several significant achievements to his credit—with at least two physics terms bearing his name. One of Oort's most notable achievements involved establishing, in 1932, how to observationally measure the amount of matter, including dark matter, in the galaxy.

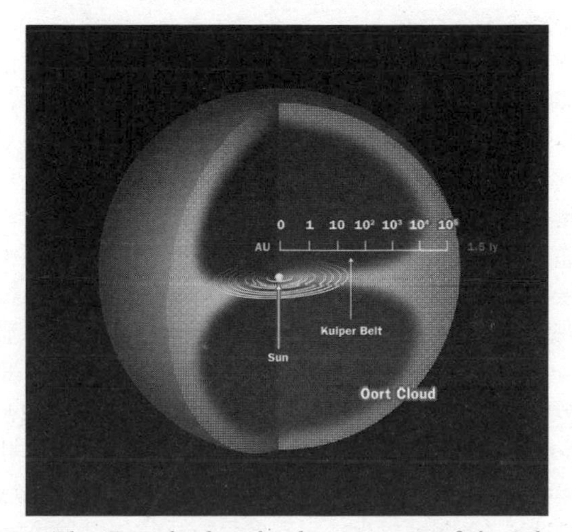

[FIGURE 15] The Oort cloud in the distant region of the Solar System, which extends from perhaps 1,000 AU to beyond 50,000 AU, lies outside the domain of the planets and the Kuiper belt.

Oort is also responsible for the speculation about what is now called the Oort cloud. In the 1930s, the Estonian astronomer Ernst Julius Öpik first suggested such a cloud's existence as the place where long-period comets originate. By 1950, Oort had both theoretical and empirical reasons to conjecture about such a spherical cloud of very distant objects. First, he observed that the long-period comets that were coming from all directions had extremely large orbits, indicating an origin much farther out than the Kuiper belt. Oort further realized that comet orbits couldn't have survived long enough to be observed today had they always been in their current paths. Comet orbits are unstable, so planetary perturbations cause them eventually to collide with the Sun or a planet or to be sent out of the Solar System altogether. Furthermore, comets "run out of steam" by passing close to the Sun too many times—the outgassing can't last forever before the objects disappear. Oort's hypothesis was that what we now call the Oort cloud replenishes the supply of fresh comets so newer ones can still be observed today.

The proposed distance of the Oort cloud is enormous. The Earth's distance from the Sun is 1 AU and that of Neptune—the most distant planet—is 30 AU. Astronomers think the Oort cloud extends from perhaps as close as 1,000 AU from the Sun to distances farther than 50,000 AU—significantly farther than anything we have so far considered. The Oort cloud reaches a good fraction of the distance from the Sun to the nearest star, Proxima Centauri—about 270,000 AU (or 4.2 light-years) away. Light from the Oort cloud would take almost a year to get to us.

The weak gravitational binding energy of objects at the distant edge of the Solar System explains why they would be vulnerable to small gravitational perturbations that can give rise to the comets we observe. Nudges can kick them out of their orbits into the inner Solar System, giving rise to long-period comets. Perturbations of these weakly bound objects might lead to short-period comets too when a planet further diverts their inward trajectories. So the Oort cloud is probably responsible for all long-period comets—such as the recently observed comet Hale-Bopp—and even some short-period ones—Halley's comet perhaps among them. Moreover, even though most Jupiter-family short-period comets probably come from the scattered disk, some of them contain isotope ratios of carbon and nitrogen that are similar to those of the long-period comets from the Oort cloud, indicating an origin there too. A final—and even more disruptive—possibility is that objects perturbed from the Oort cloud could enter the inner Solar System and collide with a planet—possibly the Earth—to create a comet strike. Later on I will return to this intriguing possibility.

Long-period comets give clues to the nature of the Oort cloud's residents. As with other comets, they contain water, methane, ethane, and carbon monoxide. But some elements of the Oort cloud might be rocky—more similar to asteroids in composition. Although called a "cloud," the reservoir of comets seems to have structure, consisting of a doughnut-shaped inner region—sometimes called the Hills cloud in deference to J. G. Hills, who proposed the separate inner region in 1981—and a spherical outer cloud of cometary nuclei in the other region, which extends much further out.

Despite its enormous size, the outer Oort cloud's total mass might be a mere five times that of the Earth. However, it most likely contains billions of low-density objects bigger than 200 kilometers across and trillions bigger than a kilometer. Models indicate that the inner region, which extends to about 20,000 AU, might contain many times this number. This inner cloud might be the source of objects that replace those lost from the more weakly bound outer Oort cloud, without which the cloud wouldn't have survived.

Because the Oort cloud is so far away, we don't have the capacity to see its icy bodies in situ. Observing small remote objects that reflect so little light from the farther reaches of the Solar System is very difficult. For objects at the enormous distances of the Oort cloud—a thousand times farther from the Sun than the Kuiper belt—observations are currently impossible. So the Oort cloud is still hypothetical in that no one has ever observed its structure or the objects it contains. Nonetheless, the Oort cloud is considered a fairly well-established component of the Solar System. The trajectories of long-period comets—coming from all directions in the sky—convincingly argue for its existence and the origin of comets in this faraway region.

The Oort cloud probably arose from the protoplanetary disk, which ultimately gave rise to much of the structure in the Solar System. Comet collisions, galactic tides, and interactions with other stars—especially in a past when those interactions were presumably more frequent—might all have contributed to the Oort cloud's formation. Objects that formed closer to the Sun in the dynamic early Solar System might have moved outward under the influence of gas giant planets to create the Oort cloud, or the population might have arisen from unstable objects in the scattered disk.

We certainly don't yet know all the answers. But in light of recent observations and theoretical work, we are learning a great deal about the outer edge of the Solar System. Perhaps we shouldn't be surprised that it is a fascinating and dynamic place.

8

THE EDGE OF THE SOLAR SYSTEM

In 1977, NASA launched the Voyager I space probe as a four-year mission to study Saturn and Jupiter. Decades later—in an incredible manifestation of persistence and robustness—it had traveled more than 125 AU away from the Earth (a distance where an emitted light signal would take the better part of a day to reach us), and was still going strong. The Voyager spacecraft and its measurements were directly accessing reaches of space that no mission has ever entered before. It's true that the data collection system based on eight-track tapes had to be tinkered with along the way, that it no longer had a working camera, and that its equipment has about a million times less memory than a smartphone does today. But the spacecraft is still operational and is currently the man-made object most distant from the Earth and the Sun.

Voyager I—despite its near obsolescence—became a hot topic in the 2013 news, when NASA announced then that on August 25, 2012, the spacecraft had entered interstellar space. The debate became quite animated—certainly in the science community, but also beyond—when news stories stated that Voyager I had reached the edge of the Solar System. On Twitter, the conversation was especially insistent and amusing. Exultant tweets over Voyager exiting the Solar System alternated with exasperated ones asking that people stop saying that Voyager was leaving. It took me a moment, but I soon realized that people weren't objecting to the repetition but were actually questioning the statement's validity. What exactly do we mean by the Solar System's end?

We've now traveled figuratively to the Oort cloud—a reasonable candidate for the edge of the Solar System—but neither Voyager nor any other spacecraft has even come close to this very distant region. Since the dark matter–meteoroid connection relies on the Oort cloud and its vicinity, let's briefly turn to the question of what exactly was meant when Voyager was said to have exited. Where is the limit of the Solar System, and why is its boundary so difficult to define?

WAS VOYAGER IN OR OUT?

The Solar System encompasses a small fraction of the size of the visible Universe, but it's nonetheless extremely big. By most reasonable measures, it encompasses the Oort cloud, which extends to at least 50,000 times the Earth–Sun distance (1 AU) and very likely twice as far—more than a light-year away. To get an idea of how distant this is, consider how long it would take for a spacecraft with current technology to reach these outer regions. A spacecraft travels about as quickly as the Earth moves around the Sun, which means it gets to a distance about equal to the circumference of the Earth's orbit in a year. According to this estimate, it would take approximately eight or nine thousand years to reach 50,000 AU, which is about a fifth the distance to the nearest star beyond our Solar System. But how many AUs big is the Solar System exactly?

The two reigning definitions give different answers, and the second one alone yields ambiguous results according to how it is delineated. In the first definition, the Solar System is the extent of the region in which the Sun's gravitational potential dominates over extra-solar gravitational influences. Defined in this way that refers to the gravitational pull of the Sun, Voyager is still in the Solar System. Really, since the Oort cloud is widely considered to be part of the Solar System, it's hard to accept that Voyager, which hasn't yet even entered the Oort cloud—and according to present estimates won't do so for at least another 300 years and won't exit it for perhaps another 30,000—is no longer in our stellar vicinity.

However, since it's unclear where the Sun's gravitational pull ends, this first definition can be fuzzy. So there is a second definition corresponding to entering interstellar space, which is characterized by where the magnetic field associated with the solar wind ends—about 15 billion kilometers away, or about 100 AU. That's so far away that radio signals emitted there take about a day to reach us. But it's quite a bit closer than the Oort cloud.

The solar wind introduced in the previous chapter consists of charged particles—electrons and protons—emitted by the Sun. These particles carry a magnetic field that streams out toward interstellar space at speeds of about 400 kilometers per second. Interstellar space by definition is the region between stars, but it's not empty. It contains cold hydrogen gas, dust, ionized gas, and some other material from exploded stars and the stellar wind of stars aside from the Sun. Eventually the solar wind and the interstellar medium abut. That region creates a cavity called the *heliosphere* and the boundary between the two regions is called the *heliopause*. Because the Solar System moves, the boundary is more tear-shaped than spherical.

Some scientists consider the heliopause to be the border between the Solar System and interstellar space. The signal that Voyager has reached the edge of the heliosphere and entered outer space would then be its encountering a decrease of charged particles from inside the heliosphere and an increase in those from outside. The particles are distinguishable because they have different energies, with high-energy charged particles originating as cosmic rays coming from distant supernovae outside the Solar System. In August 2012, Voyager data showed a sharp increase in such particles. There was also a marked decrease in low-energy particle detection. Since those originate from the Sun and the higher-energy particles come from the interstellar medium (ISM), the two measurements together were a strong indication that the probe had left the heliosphere.

However, the original criterion defining the heliopause also included the requirement of a change in strength and direction of the magnetic field to match the one that is outside the heliosphere. This definition isn't even constant over time, but depends on the Sun's "weather"—what the solar wind is doing at a given moment. It turned out that although the properties of the charged particle plasma that were measured were consistent with leaving the Solar System, they didn't satisfy the more restrictive criterion based on the magnetic field. No change in magnetic field was seen.

So although the change in plasma environment occurred on August 25, 2012, as late as March 2013 the question of whether Voyager I had entered interstellar space remained a subject of debate. Nonetheless, on September 12, 2013, NASA announced that it had. Scientists had decided the change in magnetic field was actually not required after all. They decided to go with a less restrictive criterion, which was the increase in electron density by a factor of almost 100, which is expected once outside the heliopause.

So according to the first definition—defined by the gravitational pull of the Sun—Voyager is still in the Solar System and will be so for quite some time. Even so, according to the second (newly revised) definition, Voyager has entered interstellar space. It seems that the answer to whether Voyager left the Solar System depends on which of the definitions you use.

As an amusing addendum, Voyager I carries a golden audiovisual disk with information about human society just in case an alien happens to pick it up. I suppose I'd find whatever anyone included to be pretty random, but in this case it includes greetings in English from Jimmy Carter, who was president of the United States at the time of launch, greetings in forty-nine other languages, whale sounds, and Chuck Berry's song "Johnny B. Goode." (Chuck Berry was even present for the launch.) The idea that an alien civilization, never mind our own, will be able to readily play this record in a few hundred years strikes me as slightly unlikely, as does even the idea that they should be of comparable size to us or use the necessary recording equipment—which most of us on Earth would be hard-pressed to find. I won't even bother with the translation issue or the range of sounds they are likely to appreciate should the unlikely encounter occur. But I suppose it's good to think ahead. This golden disk did have at least one positive consequence. It was the reason that Annie Druyan, who was the disk's creative director, collaborated with Carl Sagan. Even if it is most likely indecipherable to potential foreign life, the disk did contribute to a wonderful love story.

I'll leave alien visitors aside for now and turn the focus to encounters from outer space about which we can be far more certain—meteoroids that strike the Earth or at least enter our atmosphere. If Muhammad doesn't make it to the mountain, then the mountain must come to Muhammad. Which is to say that even though no one will get to the Oort cloud anytime soon, small Solar System bodies—possibly from the Oort cloud—do occasionally descend to Earth.

9

LIVING DANGEROUSLY

I recently took advantage of Harvard's spring break to visit friends in Colorado and do some work and skiing there. The Rocky Mountains are an extraordinary place to sit and think, with the nights as dazzlingly inspirational as the days. On clear dry nights, the sky is illuminated by brilliant dots of light sporadically punctuated by "shooting stars"—those tiny ancient meteoroids that disintegrate overhead. One night a friend and I stood outside the house where I was staying, mesmerized by the breathtaking expanse of luminous objects ranged densely across the sky. I had already spotted a couple of meteors before my friend and I both noticed a big one that lasted a few seconds.

Although I'm a physicist, I'm often content in such a magnificent setting to stop deliberating and simply enjoy the view. But this time I reflected on what that object was and what its trajectory might signify. The meteor—the culmination of a four and a half billion year-long story—glowed for a few seconds, implying that the visible meteoroid above might have traveled fifty to one hundred kilometers before it vaporized and disappeared. The meteoroid was probably about the same number of kilometers above us, which is why we saw it as a big arc in the sky. There it was—a thing of beauty and something we can at least partially understand. When I commented on this dust or pebble-sized object that was so wonderful to watch streak across the firmament, my friend—who is not a scientist—expressed surprise, saying he had imagined the object having a breadth of at least a mile.

The conversation rapidly swerved from calm admiration of the gorgeous sky to contemplating the damage a mile-long object careening to Earth would create. The probability of such a big dangerous object striking the Earth is small, and the likelihood that an object of any significant size would hit a populated region where it could do substantial damage is smaller still. Even so, extrapolating from the Moon's surface (too few craters survive on Earth to provide a useful guide) tells us that millions of objects bigger than a kilometer and ranging up to about a thousand kilometers across have hit the Earth over its lifetime. But most of those impacts happened billions of years ago during the Late Heavy Bombardment which, despite the name, occurred relatively soon after the Solar System was formed and before it settled into its more or less stable state.

As is essential to the survival of life, the big meteoroid hit rate is currently much lower, and has been since the bombardment episode ended. Even the recent impact in Siberia caught by dash-cams and videos—the Chelyabinsk meteoroid that burned brightly in the sky and on YouTube—was only about twenty meters across. The only recent encounter of an object as big as was envisioned by my friend was the 1994 event in which Comet Shoemaker-Levy 9's mile-sized fragments crashed into Jupiter. The initial object was bigger still—probably a few miles across before it broke into pieces. Some indication of the damage that mile-wide fragments can create was the dark cloud as big as the Earth that we could observe on Jupiter's surface. Twenty meters is big but a mile across is another thing altogether.

Bear in mind that the story of meteoroids is not solely about destruction. Some good has also come from the many meteoroids and micrometeoroids that have rained down on the Earth. Meteorites—the remaining fragments of meteoroids on Earth—might have been a source of amino acids essential to life and also of its water—another key ingredient of existence as we know it. Certainly many metals come from extraterrestrial impacts. And one can argue that humans would not have emerged without the rapid rise to dominance of mammals that occurred after a meteoroid impact—more on this in Chapter 12—killed the terrestrial dinosaurs, which, I'll grant, is not always considered to have been a good thing.

But this major mass extinction 66 million years ago is one of many stories that tie life on Earth to the rest of the Solar System. This book is about the seemingly abstract stuff such as dark matter that I study, but it is also about the Earth's relationships to its cosmic surroundings. I will now begin to explore some of what we know about asteroids and comets that have hit the Earth and the scars they've left behind. I'll also consider what might hit our planet in the future, and how we might prevent these disruptive, uninvited guests.

OUT OF THE BLUE

A phenomenon as bizarre as objects from space hitting the Earth sounds incredible and, indeed, the scientific establishment didn't initially accept the veracity of most such claims. Although people in the ancient world had believed that objects from space could reach the Earth's surface—and rural residents in more recent times were convinced of it too—the more educated classes were suspicious of the idea until well into the nineteenth century. The unschooled shepherds who had seen such objects falling from the sky knew what they'd seen, but these witnesses lacked credibility since many of those with similar backgrounds had been known to report imaginary findings as well. Even the scientists who did eventually accept that objects fell onto our planet didn't initially believe that such rocks had originated in space. They preferred for them to have an Earth-based explanation, such as the descent of material that had been ejected by volcanoes.

The arrival of meteorites from outer space became part of established thinking only in June 1794, after a fortuitous fall of stones over the Academy in Siena—where many educated Italians and British tourists could witness the event firsthand. The dramatic phenomenon began with a high, dark cloud that emitted smoke, sparks, and slow-moving, red lightning that was followed by stones that rained down to the ground. The Abbe Ambrogio Soldani in Siena found the fallen material interesting enough to gather eyewitness accounts and to send a sample to a chemist living in Naples—Guglielmo Thomson—the alias of a relocated William Thomson who had fled Oxford in disgrace over his activities with a servant boy. Thomson's careful investigation indicated an extraterrestrial origin for the object, offering a more consistent explanation than the far-fetched proposals then in circulation involving a lunar origin or lightning hitting dust and one that was also better than the more credible competing proposition that it had originated in the then-active Vesuvius. Taking the source to be volcanic activity was understandable in that Vesuvius had coincidentally erupted only 18 hours earlier. But Vesuvius is located some 320 kilometers away and in the wrong direction, which ruled it out as an explanation.

The case for a meteoroid origin was finally settled by the chemist Edward Howard with the assistance of the French nobleman and scientist Jacques-Louis, Comte de Bournon, who had been exiled to London during the French Revolution in 1800. Howard and the Comte analyzed a meteorite that had fallen near Benares in India. They uncovered an amount of nickel that was much greater than would be expected on the Earth's surface as well as stony materials that had been fused by high pressure. The chemical analyses that Thomson, Howard, and the Comte performed were precisely the sort of thing that the German scientist Ernst Florens Friedrich Chladni had suggested to confirm his own hypothesis that such objects hit the Earth at too great a speed to be consistent with other proposed explanations. In fact, the Siena fall occurred only two months after the publication of Chladni's book, *On the Origin of Ironmasses,* which had—alas—received negative reviews and an unfavorable response before the Berlin newspapers eventually got round to reporting the Siena fall two years after it occurred.

More widely read in England was the short book that Edward King, a fellow of the Royal Society in England, published that year. King's book reviewed the Siena event and a lot of Chladni's book too.

The case in England for meteorites had been further solidified even earlier as well when a 56-pound stone fell on December 13, 1795 at Wold Cottage in Yorkshire. With an increased appreciation of the methods of chemistry—just recently separated from alchemy—and with so much firsthand evidence, meteorites were finally recognized in the nineteenth century for what they were. Many objects with bona fide extraterrestrial credentials have fallen to Earth since that time.

MORE RECENT EVENTS

Headlines about meteoroids and meteorites are pretty much guaranteed to spark our interest. But even while avidly following these remarkable events, we shouldn't forget that today we generally live in equilibrium with the Solar System, and we rarely encounter dramatic disruptions. Almost all meteoroids are small enough to disintegrate in the upper atmosphere where most of their solid material gets vaporized. Bigger objects arrive only infrequently. But small objects do visit us, and they do so all the time. Mostly micrometeoroids enter the atmosphere, and these particles are so small they don't even burn up. Though less frequent, millimeter-sized objects enter the Earth's environs pretty often too—perhaps once every 30 seconds—and those burn up with no significant consequences. Objects bigger than about two or three centimeters partially burn up in the atmosphere so fragments of these might make it to the ground, but those will be too small to be significant.

But every few thousand years, an explosion caused by a big object low in the atmosphere might occur. The largest such event ever recorded occurred in 1908 in Tunguska, Russia. Even without any surface impact, an explosion in the atmosphere can produce noticeable consequences on Earth. This particular asteroid or comet—we often don't know which—burst in the sky near the Tunguska River in the forests of Siberia. The power of this roughly 50-meter-sized *bolide*—an object from space that disintegrates in the atmosphere—was the equivalent of about 10 to 15 megatons of TNT—1,000 times bigger than the Hiroshima explosion but not quite as big as the largest nuclear bomb ever detonated. The explosion destroyed 2,000 square kilometers of forest and produced a shock wave that would have measured about 5.0 on the Richter scale. Notably, the trees at what was almost certainly ground zero were left standing, while the surrounding ones were smashed down flat. The size of the zone of the upright trees—and the absence of a crater—meant that the impactor likely disintegrated about six to ten kilometers above the ground.

Risk estimates vary, in part because of the changing estimate of the Tunguska object's size—which have ranged from thirty to seventy meters. Something with size in this interval might hit at a rate ranging from once every few hundred years to once in every couple of thousand. Even so, most of the meteoroids that do hit or come near Earth approach relatively unpopulated regions, since the distribution of dense population centers is sparse.

The Tunguska meteoroid was no exception in this regard. It exploded over an unsettled area in Siberia where the closest trading station was seventy kilometers away and the nearest village—Nizhne-Karelinsk—was farther away still. Even so, the blast was sufficiently strong in this not-so-closeby village to knock out windows and topple pedestrians. Villagers had to turn away from the blinding flash in the sky. Twenty years after the explosion, scientists returned to the region to learn that some local herdsmen had experienced noise and shock trauma, with two of them actually killed by the impact. The consequences for the animal world were devastating, with perhaps a thousand reindeer killed by the fire that the impact left in its wake.

The event influenced a much larger region as well. The blast was heard by people living at a distance as far away as France is wide, and barometric pressure changed all over the Earth. The wave from the explosion circumnavigated the globe three times. In fact, many of the destructive consequences of the larger and better studied Chicxulub impact I'll soon get to—the one that killed the dinosaurs—happened after the Tunguska event too, with winds, fires, climate change, and the disappearance of about half the ozone in the atmosphere.

Yet because the meteoroid exploded in a remote and unpopulated region and in a time and place where mass communication was minimal, most people barely paid attention to this tremendous blast until decades later, when an investigation finally revealed the full extent of the devastation. Tunguska was remote, and was isolated even further by the First World War and the Russian Revolution. Had the explosion happened a mere hour earlier or later, it might have hit a major population center, in which case atmospheric effects or an ocean tsunami would probably have killed thousands of people. Had this been true, the impact would have shaped not only the surface of the globe, but the history of the twentieth century—most likely with politics and science unfolding quite differently as a consequence.

Several smaller, but nonetheless newsworthy, celestial visitors have come to Earth in the hundred years since the Tunguska explosion. Though poorly documented, a bolide that burst in the atmosphere above the Amazon in Brazil in 1930 might have been among the larger ones. The net energy released was less than that of the Tunguska event, with estimates varying from 1/100 to 1/2 as big. Even so, the meteoroid mass was more than 1,000 tons, and might have been as massive as 25,000 tons—yielding an energy of about 100 kilotons of TNT. Risk estimates vary, but objects between 10 to 30 meters in size might hit at a rate ranging from roughly once a decade to once every few hundred centuries. The rate estimate strongly depends on the exact size of the object. An uncertainty in size by a factor of two can lead to estimates varying by up to a factor of ten.

A bolide that was similar in size to the one over the Amazon exploded about 15 kilometers above Spain a couple of years later, releasing the equivalent of about 200 kilotons of TNT. A number of explosions occurred in the next 50 years or so, though none were even as big as the Brazil event and I won't list them all. One event of note was the Vela Incident of 1979, which occurred between the South Atlantic and the Indian Ocean and was named after the U.S. Vela defense satellite that observed it. Though initially considered a plausible meteoroid candidate, people now attribute it to a nuclear blast that was detonated here on Earth.

Of course sensors detect actual bolides too. Department of Defense infrared sensors and Department of Energy visible wavelength sensors picked up the signal from a 5 to 15 meter-wide meteoroid that exploded on February 1, 1994 over the Pacific Ocean near the Marshall Islands. Two fishermen off the coast of Kosrae, Micronesia, a few hundred kilometers from the impact, detected it too. Another even more recent explosion of a 10-meter-wide object occurred in 2002 over the Mediterranean Sea between Greece and Libya, releasing the energy equivalent of about 25 kilotons of TNT. More recent still was the October 8, 2009 event near Bone, Indonesia, which probably also originated from an object about 10 meters in diameter and which released up to 50 kilotons of energy.

Errant comets or asteroids can both be a source of meteoroids. The trajectories of distant comets are difficult to predict, but sufficiently large asteroids can be detected well before they arrive. An asteroid that made impact in 2008 in Sudan was significant in this respect. On October 6 of that year scientists calculated that the asteroid they had just found was about to hit the Earth the following morning. And indeed it did. It wasn't a major impact and no one lived in the vicinity. But it demonstrated that some impacts can be predicted, though how much advance notice we'll receive will depend on our detection sensitivity relative to the object's size and speed.

The most recent newsworthy event was the February 15, 2013 Chelyabinsk meteor, imprinted not just in pictures but also in living memories. This bolide explosion 20 to 50 kilometers above the southern Ural region of Russia generated about 500 kilotons of TNT's worth of energy—most of which was absorbed by the atmosphere—though a shock wave carrying some of the energy hit the Earth several minutes later too. The event was triggered by an asteroid of about 15 to 20 meters across that weighed about 13,000 tons and descended with an estimated speed of 18 km/sec—about sixty times the speed of sound. Not only did people see the explosion—they felt the heat from its atmospheric entry too.

About 1,500 people were injured by the event—but mostly from secondary consequences such as blown-out panes of glass. The number of people affected was inflated by the many witnesses who had gone to their windows to see the blinding flash that—traveling at the speed of light—was the first sign of something odd. In an unfortunate twist—worthy of a good horror movie—the light in the sky had lured people to precarious locations right before the impact of the shock waves hit and did most of the damage.

Adding to the media frenzy, at the time the meteoroid hit, news reports had warned about a different asteroid that also appeared to be approaching Earth. The Chelyabinsk meteoroid snuck up undetected, while this other 30-meter object—which made its closest approach about 16 hours later—never made it to the Earth's atmosphere. Many people speculated that the two asteroids had a common origin, but according to follow-up studies, this probably wasn't true.

NEAR-EARTH OBJECTS

Like the predicted asteroid of February 2013, a number of close approaches by objects that never actually hit or entered the atmosphere have attracted plenty of attention. Other objects do arrive to Earth—but even among these, the overwhelming majority of them are harmless. Nonetheless, past collisions have influenced geology and biology on the planet, and could well do so again in the future. With the increasing appreciation of asteroids and the (possibly exaggerated) awareness of their potential danger, the search for asteroids with the potential to cross the Earth's orbit has intensified.

The most frequent encounters—though not necessarily the biggest—come from what are known as *near-Earth objects (NEOs)*—stuff that is pretty close to Earth, with closest approach to the Sun no farther than 30 percent more than the Earth-Sun distance. About ten thousand near-Earth asteroids (NEAs) and a smaller number of comets meet this criterion, as do some large meteoroids within tracking range—and technically some solar-orbiting spacecraft too.

NEAs are divided into several categories. (See Figure 16.) The bodies that enter the Earth's domain and come close without actually intersecting our orbit are called *Amors*—named after the 1932 asteroid that came within 16 million kilometers, or a mere 0.11AU. Although they don't currently cross our path, the potential fear is that Jupiter or Mars-induced perturbations could increase these objects' eccentricities so that they do ultimately intersect our orbit. The *Apollos*—again named after a particular asteroid—are ones that currently cross Earth's orbit in the radial direction, though they can be above or below our ecliptic—the apparent path of the Sun in the sky that denotes the plane of the Earth's orbit—so that they usually don't actually intersect. The path can however change over time—again possibly making it deviate into a dangerous zone. A second category of Earth-crossing asteroids—distinguished from Apollos by their orbital domains, which are smaller than the Earth's—are known as *Atens.* The Aten family is again named after an asteroid of its type. The final NEA category is *Atiras*—those asteroids whose orbital domains lie entirely within that of the Earth. They are hard to find, so only a few are known.

NEAs don't last all that long on geological and cosmological scales. They hang around for only a few million years before they are thrown out of the Solar System or collide with the Sun or a planet. That means in order to populate the region close to the Earth's orbit, new asteroids have to be in continuous supply. They are probably created by Jupiter's perturbations to the asteroid belt.

Most of the NEAs are stony asteroids, but there are a fair number of carbon-containing carbonaceous asteroids as well. The only

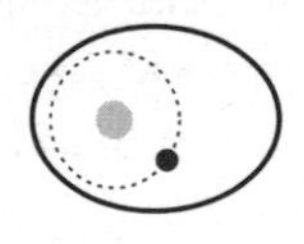

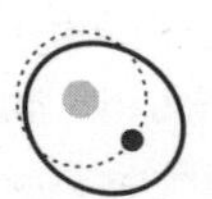

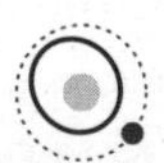

[FIGURE 16] The four categories of Near-Earth Asteroids. Amors' orbits lies between that of the Earth and of Mars. Apollos' and Atens' paths cross the Earth's orbit, but can extend beyond it for some fraction of the orbital period. Apollos have semi-major axes greater than the Earth's, whereas Atens have semi-major axes that are smaller. The orbits of Atiras lie entirely within that of the Earth.

ones bigger than 10 kilometers wide are Amors—which don't cross our path at present. However, there are a fair number of Apollos bigger than five kilometers across—certainly big enough to do a fair bit of damage should the trajectory prove to be infortuitous. The biggest NEA at 32 kilometers across is Ganymed, which is the German spelling of the Trojan prince whom the English call Ganymede. Ganymede, one of Jupiter's moons, is a completely different object but also wins a size contest as it is the largest moon in the Solar System.

NEAs constitute another research area that has ripened in the last 50 years. Earlier on, no one even took the idea of impacts very seriously. Now people around the globe have begun to catalog and track NEAs wherever possible. Even on my recent visit to the Canary Islands, when I visited the Tenerife telescope, I found the director with a dozen students who were examining data to try to find them. The small old telescope there isn't state-of-the-art, but I was impressed to see the motivated students and their appreciation of the methods used to search.

Today's more advanced telescopes seek asteroids through the use of charge-coupled devices, which employ semiconductors to turn photons into charged electrons, leaving signals that pinpoint where the photons had hit. Automated readout systems have also helped escalate the discovery rate. The website of the International Astronomical Union's Minor Planet Center of the Harvard Smithsonian Center for Astrophysics, http://www.minorplanetcenter.net/, reports the latest numbers of minor planets, comets, and near approaches that have been found.

For obvious reasons, the orbits that are near that of the Earth receive the most attention. The United States and the European Union collaborate on scanning for these in an enterprise called Spaceguard—coined as a shout-out to Arthur C. Clarke's science fiction novel *Rendezvous with Rama*. The task of the first Spaceguard program was determined in a 1992 US Congressional survey report, which led to a mandate to categorize within a decade most near-Earth objects that are bigger than a kilometer. A kilometer is big—

bigger than the smallest object with the potential to do harm—but was chosen because kilometer-sized objects can be found more readily and are sufficiently big to do world-wide damage. Fortunately, of the kilometer-sized objects we know about, most orbit between Mars and Jupiter in the asteroid belt. Until they change orbits to become NEOs, they certainly pose no threat.

By careful use of observations, projected orbits, and computer simulations, astronomers achieved Spaceguard's goal of identifying most kilometer-sized NEOs in 2009, almost on schedule. Current findings suggest about 940 near-Earth asteroids of a kilometer or more in size. A committee convened by the National Academy of Sciences determined that even with uncertainties accounted for, this number is pretty accurate, with the total number expected to be less than 1100. These searches have also helped identify about 100,000 asteroids and approximately 10,000 NEAs that are smaller than a kilometer.

Most of the larger NEAs that were the target of the Spaceguard mandate come from the inner and central regions of the asteroid belt. The National Academy committee determined that about 20 percent of the orbits for which they have statistics pass within 0.05 AU of Earth. They called these more precariously located ones "potentially hazardous NEOs." The Academy also determined that none of these objects pose a threat within the coming century, which is of course welcome news. The result is not all that surprising however since one-kilometer objects are expected to strike the Earth no more than once every few hundred thousand years.

In fact, there is only one known NEO with any measurable probability of hitting the Earth and doing damage in the near future. But the probability that it will come close is a mere 0.3 percent, and even that isn't projected to happen until 2880. We are almost certainly very safe—at least for the time being—even with all the uncertainties accounted for. Some astronomers had earlier on raised concerns about a different asteroid—the demonically named, 300 meter-wide Apophis, which they had projected would miss the Earth in its close approach in 2029 but potentially return for an impact in 2036 or 2037.

This was supposed to follow its passing through a "gravitational keyhole," which they thought might have the potential to send it speeding off in our direction. However, further calculations revealed this to be a false alarm. Neither Apophis nor any known object should hit us in the foreseeable future.

But before breathing too big a sigh of relief, keep in mind that we still have smaller objects to worry about. Though objects smaller than the kilometer-size that Spaceguard originally targeted would do less damage, they should come close or strike more frequently. So Spaceguard was extended by a 2005 congressional mandate* to encourage the United States to track, catalog, and characterize at least 90 percent of potentially dangerous near-Earth objects bigger than 140 meters across. They almost certainly won't find something truly catastrophic, but the catalog is nonetheless a worthwhile goal.

ASSESSING RISK

Clearly asteroids sometimes come close. Encounters will undoubtedly occur, but their expected frequency and magnitude remain subjects of debate. Whether or not something will hit and do damage on time scales that we should care about is not yet an entirely settled question.

Should we worry? It's all a matter of scale, cost, our anxiety threshold, the decisions societies make about what is important, and what we think we can control. The physics in this book primarily concerns phenomena that occur over million or even billion year time scales. The model that I worked on, which the next part of the book will describe, might account for a 30 to 35 million year periodicity for large (few kilometer or so) meteoroid hits. None of these are time scales that are particularly worrisome or relevant to humanity. People have much more pressing concerns.

However, even if it's a bit of a digression, I couldn't very well write

* The George E. Brown, Jr. Near-Earth Object Survey section of the NASA Authorization Act of 2005 (Public Law 109-155)

a book that touches on meteoroid strikes without at least giving some feel for established scientists' conclusions about their potential impact on our world. The topic comes up in the news and conversation enough that it won't hurt to share some current estimates. The projections are relevant to governments too when they consider how important asteroid detection and deflection would be.

In accordance with Congress' 2008 Consolidated Appropriations Act, NASA asked the National Research Council of the prestigious National Academy of Sciences to study near-Earth objects. The goal was not to address any of the abstract impact questions, but to evaluate the risk posed by errant asteroids and whether something could be done to mitigate that risk.

The participants focused their study on smaller NEOs, which hit much more frequently and can potentially be diverted. Comets in short-period orbits are similar to asteroids in their trajectories, so they can be detected in a similar manner. Long-period comets, on the other hand, are virtually impossible to see in advance. They are also less likely to be in the equatorial plane of the Earth's orbit—they come from all directions—so finding them is more difficult. In any case, though some of the recently observed events might have originated in comets, comets reach the Earth's vicinity far less frequently. And it would be pretty much impossible to identify long-period comets in time to do anything—even if technological advances do eventually enable us to deflect asteroids. Since there is virtually no way at this point to make a complete catalog of hazardous long-period comets, current surveys focus only on asteroids and short-period comets.

But long-period comets—or at least those comets that arise from the outer Solar System—will be the ones of interest to us later. Objects arising from the outer Solar System are far more weakly bound, so disturbances—gravitational or otherwise—can more readily send one out of its orbit and into the inner Solar System or else out of the Solar System altogether. Though not the subject of the National Academy's mitigation studies, they can still be the subject of scientific investigations.

THE SCIENTISTS' CONCLUSIONS

In 2010, the National Academy of Sciences presented their results on asteroids and the threats they pose in a document titled *Defending Planet Earth: Near-Earth Object Surveys and Hazard Mitigation Strategies*. I'll present a few of the document's more interesting conclusions, reproduce a few of the tables and charts that best summarize them, and add a few words and comments to explain what they mean.

When interpreting the numbers, remember to factor in the relatively low density of heavily populated urban areas, which the Global Urban Mapping Project estimates as approximately three percent. Though any destruction would of course be unwanted, the scariest threat would be to an urban area. The low density of cities on the Earth's surface tells us that the frequency of any relatively small objects hitting and causing significant damage is about 30 times lower than their frequency of impact. For example, if a five to ten-meter-sized object is predicted to hit roughly once every century, something of this size would be expected to hit cities only about once in three millennia.

We should also take note of the large uncertainties in almost all of the projections, which scientists can estimate at best to within a factor of ten. One reason for the many news stories about distant threats that never materialize is that even for particular objects of particular sizes, minor errors in measuring trajectories can make a big difference to the predicted likelihood of a hit. We also don't fully understand the effects and the damage that even known big objects can cause. Even with such uncertainties, the results from the National Academy study are fairly reliable and useful. So with the allowance of some degree of uncertainty, let's now explore these fascinating state-of-the-art (circa 2010) statistics.

My favorite table is in Figure 17. According to these results, an average of 91 deaths by asteroid occur every year. Although asteroids are far behind most catastrophic causes of death—rates are comparable to those from fatal wheelchair-associated accidents (not listed)—the number 91 in the table next to asteroids is a little surprisingly and uncomfortably high. It is also ridiculously precise given all the

Expected Fatalities per Year, Worldwide, from a Variety of Causes

CAUSE	EXPECTED DEATHS PER YEAR
Shark attacks	3-7
Asteroids	91
Earthquakes	36,000
Malaria	1,000,000
Traffic accidents	1,200,000
Air pollution	2,000,000
HIV/AIDS	2,100,000
Tobacco	5,000,000

[FIGURE 17] NAS statistics for average worldwide fatalities per year from a variety of causes. Statistics based on data, models, and projections.

uncertainties. Clearly, 91 deaths by asteroid do not occur every year. In fact we know of only a few such deaths in recorded history. The large number is deceptively high because it includes enormous hits that are predicted to occur only very rarely. Here's an edifying graph (Figure 18) that helps explain.

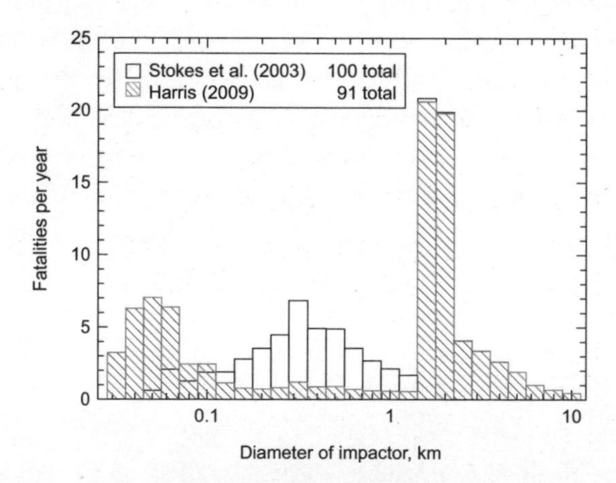

[FIGURE 18] NAS estimates of average yearly fatalities caused by impacts from asteroids of various sizes based on data from an 85 percent complete Spaceguard Survey. This plot uses the now-revised near-Earth object size distribution and updated estimates for threats from tsunamis and airbursts. Older estimates are also shown for comparison.

What this graph tells you is that the majority of the number quoted above comes from larger objects, which are predicted to occur extremely infrequently. That's the spike at a few kilometers. Such events are the "black swans" of asteroid hits. If you restrict your attention to objects less than 10 meters in size, the number goes down to less than a few per year, which is still probably on the high side. So what are the expectations for how often objects of different sizes will actually hit? Here's one more graph (Figure 19) that should help. This one is a little busier, but bear with me. It's actually a great summary of our current understanding.

Though harder to read, this plot contains a lot of information. It uses what is known as a logarithmic scale. This means that changes in size correspond to much greater time frame variations than you

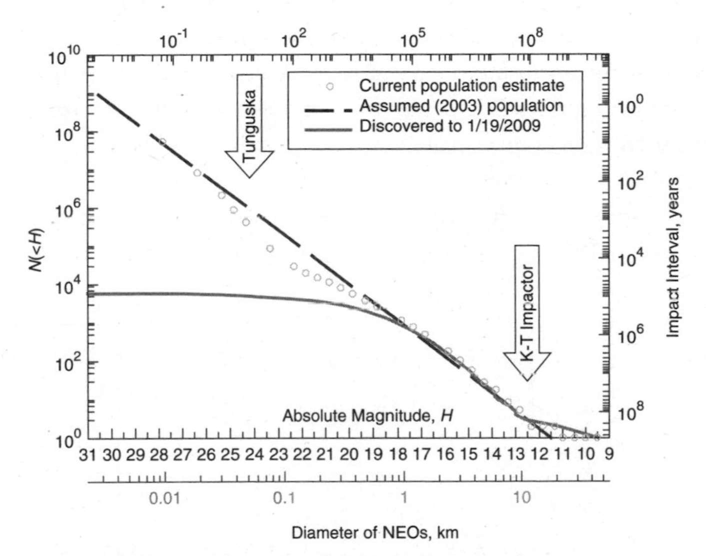

[FIGURE 19] Estimated number (left vertical axis) and approximate time between impacts (right-hand axis) of near-Earth objects as a function of diameter measured in kilometers. The top axis gives the expected impact energy in megatons of TNT for an object of a given size, assuming it was moving at 20 kilometers per second on impact. Also shown near the lower horizontal axis is a quantity related to the object's intrinsic brightness. The different curves are based on older (solid) and newer (circles) estimates. The lower curve represents the number discovered prior to 2009.

might have in mind. For example, a 10 meter sized object might arrive once per decade whereas a 25 meter object might impact the Earth once every 200 years. It also means that small changes in measured values could significantly affect predictions.

The top axis of this graph refers to how much energy an object of a given size will release, assuming it is traveling at 20 km/sec, as measured in megatons. So, for example, a 25 meter object would release about one megaton. The plot also tells how many objects of various sizes can be expected, and how bright they are likely to be—also related to how easy they are to track and find. Despite the larger number of smaller asteroids, their diminutiveness and their consequently lower brightness makes them more challenging to discover.

The estimates of frequency of these events would be, for example, a 500 meter-sized object about every 100 millennia, kilometer-sized objects perhaps once in 500,000 years, and five kilometer objects on a scale closer to 20 million years. The graph also tells you that a dinosaur-killing-sized impactor of about 10 kilometers in size is expected only about once every ten to one hundred million years.

If you are solely interested in how often strikes occur, the information is clearer in the simpler plot in Figure 20. Notice that the vertical axis has the fewest years at the top and the most at the bottom,

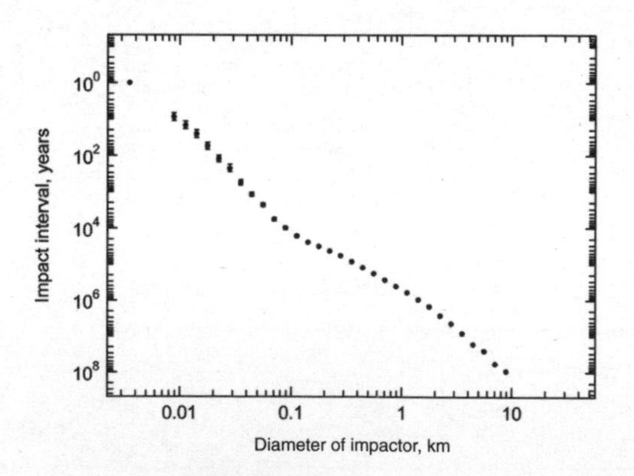

[FIGURE 20] Average numbers of years between impacts on Earth of near-Earth objects of sizes ranging from about three meters to roughly nine kilometers across.

so big impacts occur much less frequently than small ones. Notice also the exponential numbers in the vertical column that tell the number of times 10 is multiplied by itself. For example, 10^1 is ten, 10^2 is one hundred and 10^0 is one.

Finally, to get some idea of the extent of the danger from objects of various sizes, I'll present one final diagram from the Academy study in Figure 21. This table tells us that for something a few kilometers in diameter, the entire globe would be affected. Large meteoroid hits don't occur nearly as often as other natural disasters, so they almost certainly don't pose any immediate threat. But if they were to occur, their impact in terms of energy and severity would be devastating. The table also shows, for example, that something 300 meters across might hit the Earth every hundred thousand years. This could increase the sulfur in the atmosphere to levels comparable to those caused by Krakatau, damaging life or at least agriculture on a large part of the planet. And, like the earlier plots, it shows us that a Tunguska-sized airburst might occur about once every thousand years. The full contours of any of these disaster scenarios would of course depend on the size and the location of the hit.

Approximate Average Impact Interval and Impact Energy for Near-Earth Object

TYPE OF EVENT	CHARACTERISTIC DIAMETER OF IMPACTING OBJECT	APPROXIMATE IMPACT ENERGY (MT)	APPROXIMATE AVERAGE IMPACT INTERVAL (YRS)
Airburst	25 m	1	200
Local scale	50 m	10	2,000
Regional scale	140m	300	30,000
Continental scale	300m	2,000	100,000
Below global catastrophe threshold	600m	20,000	200,000
Possible global catastrophe	1km	100,000	700,000
Above global catastrophe threshold	5km	10 million	30 million
Mass extinction	10km	100 million	100 million

[FIGURE 21] Approximate average impact interval and impact energy for near-Earth objects of various sizes. Keep in mind that these quantities depend on the impactor's velocity and physical and chemical characteristics.

WHAT TO DO

So what should we conclude from all this? First of all, it's fascinating that all these objects in space orbit in the same general vicinity. We think of Earth as special and of course want to protect it. But in a larger picture it is just one of the inner planets in a Solar System orbiting around one particular star. However, even as we acknowledge the proximity of our neighbors, the second point to take away is that an asteroid isn't the greatest threat to humanity. Impacts might occur and might even do some damage, but people aren't actually in much imminent danger—at least in this regard.

Even so, the question of what to do should something dangerous appear is bound to come up. We'd feel pretty silly if we were watching an object on a dangerous Earth-bound trajectory for a few years but were impotent to do anything to improve our fate. The lack of grave danger doesn't mean we should be entirely helpless to protect against any damage a meteoroid might do or that we should never think about mitigation.

Not surprisingly, a number of people have thought about the problem and many proposals—though no actual devices—for dealing with dangerous objects from space are under consideration. The two basic strategies are destruction or deflection. Destruction *per se* isn't necessarily a great idea. If you blow up something in danger of hitting the Earth into a lot of pieces of rock hurtling in the same direction, you will most likely increase the odds of a hit. Though the damage from any particular piece should be less, a strategy that doesn't encourage a greater number of hits would be better.

So deflection is probably the more sensible approach. The most efficient deflection strategies involve increasing or decreasing the speed of an incoming object—not a sideways push. The Earth is pretty small and moves pretty quickly around the Sun—at about 30 km/sec. According to the direction from which the incoming object approaches us, changing its path so that it arrives earlier or later by a mere seven minutes—the time it takes for the Earth to move a distance of its radius—can be the difference between a collision and an exciting but harmless flyby. That's not a huge change in orbit. If something is detected early enough—perhaps a few years in advance—even a small change in velocity should suffice.

None of the suggestions for deflection or destruction would save us from an object bigger than several kilometers in size that is capable of doing global damage. Fortunately, such an impact probably won't occur for at least another million years. For smaller objects for which we can in principle save ourselves, the most effective deflectors would be nuclear explosives, which could perhaps prevent an impact for something as big as a kilometer across. However, laws forbid nuclear explosions in space, at least for now, so that technology is not being developed. Also possible, though not nearly as powerful, would be a collision of some object with an incoming asteroid so that it transfers kinetic energy, which is its energy of motion. If there is sufficient advanced notice, and especially with the possibility of multiple impacts, this strategy might work for incoming objects several hundred meters across. Other suggestions for deflectors include solar panels, satellites acting as gravitational tug boats, rocket engines—anything that could potentially create enough force. Technology along these lines might ultimately be effective for objects as big as a hundred meters, but only with a few decades advanced warning. All of these methods (and asteroids themselves) require further study so it is likely too soon to say with certainty what will work.

Such proposals, though interesting and worthy of consideration, are currently only possible visions for the future. No such technology currently exists. However, one project, the Asteroid Impact and Deflection Assessment mission—designed to test the feasibility of kinematic impact on an asteroid—is relatively far along in its planning. Also in the works is another related project—the Asteroid Redirect Mission—which would deflect an asteroid or piece of one into orbit around the Moon and perhaps host an astronaut visit later on. However, no actual construction on any of these projects has begun.

Some people would argue against building anti-asteroid technology on the grounds that it could be harmful in a broader sense. Some are afraid, for example, that such technology would be used for military purposes rather than for saving the Earth—though I find this highly unlikely in light of the required long lead time for any mitigation device to be effective. Others raise the potential psychological and sociological danger of finding an asteroid on an Earth-intercept trajectory when it is too late or beyond our technological ability to do anything about it anyway—which strikes me as a delaying tactic that can be used against a lot of potentially constructive proposals.

Such spurious concerns aside, we can nevertheless ask if we should make any preparations and, if so, when. This is really a cost and resource allocation issue. The International Academy of Astronautics holds meetings to address precisely these sort of questions and identify the best strategy. A colleague who attended the 2013 Planetary Defense Conference in Flagstaff, Arizona told me about an exercise in which the attendees were supposed to consider a fake approaching asteroid and ask themselves how best to address the simulated threat. They were asked to respond to questions such as "how to deal with uncertainties in its size and orbit that get updated over time," "when is it appropriate to act," "at which point should you call the President?," (the meeting was in America after all) "when is the time to evacuate a region," and "when would you launch a nuclear missile to deter a potential tragedy?" These questions—though at some level to my mind rather entertaining—make clear that even well-intentioned, well-informed astronomers can have very different attitudes and responses to an object approaching from space.

I hope I've convinced you that such threats are not overly pressing, even if some potential damage is possible. Though it's possible that an infelicitously directed one could hit and wipe out a major population center, the odds of that happening anytime in the foreseeable future are extremely remote. The scientist in me is all for cataloguing and understanding the trajectories of as many objects as we can. And the geek in me thinks a spacecraft that can escort a potentially dangerous NEO into a safe orbit that won't ever hit the Earth would be very cool. But really, no one knows for certain how best to proceed.

The ultimate issues for society, as with all scientific and engineering efforts, are what we value, what we learn, and what the ancillary benefits might be. You can now consider yourself armed with some basic facts if and when you choose to weigh in with an opinion. Current numbers help, but they are not complete. As with many policy choices, we need to combine educated guesses with practical considerations and moral imperatives. My feeling is that even without any threat, the science is sufficiently interesting to merit the relatively minor investment needed to find more asteroids and study them further. But only time will tell what society—and private industry—ultimately decides.

10

SHOCK AND AWE

On a recent trip to Greece, I was occasionally humbled by the impressive English vocabulary of some of the locals I met there, who would at times say a word that I—a native speaker—might hesitate to use. I remarked on this when someone spoke the word *eponymous,* which my conversational partner reminded me originated as Greek. As did many of our words, of course.

The word *crater* is among them. Apparently the ancient Greeks, though big wine drinkers, appreciated moderation too. Unless revelry was in store, wine would be mixed with three times its volume in water, and a *krater* was the designated blending vessel. A krater has a large round opening—a bit similar in shape to the huge gaping regions on Earth and the Moon that share their name. But the geological feature with a similar name can run up to 200 kilometers across, and the surrounding disturbed region can be even larger.

Some craters are formed here on Earth by volcanoes, without any outside assistance. On Tenerife on the Canary Islands, for example, you can see a few fantastic craters in the big lava field of the El Teide volcano—evidence of turmoil beneath the Earth's surface that occasionally bubbles out. This is also where I learned that *caldera* was the Spanish word for "cauldron," teaching me that the term we use for a volcanic depression has an origin similar to that of the word "crater."

Impact craters, on the other hand, are formed in isolation and—more significantly—only with an extraterrestrial contribution.

Most meteoroid hits—including all the big ones—happened well before people were around to watch, never mind record, them. An impact crater is the remarkable calling card that a speeding meteoroid that descended to Earth leaves in its wake. The craters or depressions and the material in and around them often are the sole surviving records of the disorderly visitors that wreaked havoc upon arrival. The scars, rock types, and chemical abundances buried in the debris provide our most reliable information about these long-ago events.

Impact craters serve as extraordinary evidence of the Earth's lasting connection to its environment—namely the Solar System. Understanding the formation, shape, and characteristics of impact craters helps us determine how often rocks of various sizes hit the Earth, as well as debate the possible role of meteoroids in extinctions in a more informed manner. In this chapter I'll explain why and how those awe-inspiring craters formed in the first place—and what distinguishes impact craters from the terrestrially induced depressions from volcanoes. I'll also comment on the list of objects that struck with enough force to make a lasting impression, which are nicely cataloged in the Earth Impact Data Base you can pull up on the Internet. These observations will be critical later on when I consider dark matter's role in triggering meteoroid impacts.

THE METEOR CRATER

Before diving into impact crater formation and the full list of those here on Earth, let's take a moment to reflect on the first one found—one of the earliest discoveries to tie together objects from the sky with the surface of the Earth. (See Figure 22.) Although the name is a little off—remember, "meteor" is the streak in the air—the Meteor Crater was at least formed by a meteoroid, as, by definition, are all impact craters. This particular crater is located near Flagstaff, Arizona. Its name correlates with a nearby post office, in accordance with a meteoroid naming convention. Theodore Roosevelt established

[FIGURE 22] The roughly kilometer-wide Meteor (Barringer) Crater located in Arizona. (Aerial image courtesy of D. Roddy.)

the post office in 1906 when his friend Daniel Barringer, a mining engineer and businessman, started investigating the mysterious crater's contents and origin. Geologists were initially skeptical of his proposal, but Barringer ultimately showed that the crater originated in a meteoroid. The depression is also known as the Barringer Crater in recognition of his contribution.

Though bigger impact structures exist, the crater is among the largest in the United States—measuring some 1,200 meters across and 170 meters deep, with a rim that rises about 45 meters. It is about 50,000 years old, and you can see it right on the surface of the Earth. If it wasn't obvious from a map, you can tell the crater is in America since, like many things American, it is privately owned. The Barringer family holds the title through the Barringer Crater Company and currently charge 16 dollars to see it. The ownership was secured in 1903, when Daniel Barringer staked a claim along with the mathematician and physicist Benjamin Chew Tilghman, which was soon afterward signed by the president. The company staking the claim—the Standard Iron Company—was given permission to mine with a 640-acre land patent.

Because it is private property, the crater cannot be part of the national park system. Only federally owned land can house a national monument, so it is merely a national natural landmark. The good part of that is that it doesn't get closed when the government shuts down, as occurred in 2013 when I started writing this chapter. The other good thing about private ownership is that the Barringers have a vested interest in preserving the crater, and it is indeed considered the world's best-preserved meteor impact site—though its relatively recent origin helps a lot too.

The meteorite associated with the crater is called the Diablo meteorite, named after the ghost town of Canyon Diablo, located along the canyon sporting the same name. The associated 50-meter-wide meteoroid, composed of almost pure iron and nickel, probably hit the ground at about 13 kilometers per second, generating at least two megatons of TNT worth of energy—a few times as much as Chelyabinsk or roughly the energy of a hydrogen bomb. Most of the initial object was vaporized, making fragments difficult to find. Pieces that have been located are on display in the museum there, and some are even for sale.

The paucity of fragments made it difficult at first to ascertain that the crater was in fact formed by an extraterrestrial object, rather than by a volcano, as the European settlers who first ran across it in the nineteenth century had assumed. This wasn't an unreasonable hypothesis at the time given how exotic an extraterrestrial explanation must have seemed, and how misleading the proximity of the San Francisco volcano field—only 40 miles west—must have been.

In an illuminating story of science gone wrong—and only later put right—the U.S. Geological Survey's chief geologist, Grove Karl Gilbert, made his official determination that it was a volcano in 1891. Gilbert had heard about the crater from the Philadelphia mineral dealer Arthur Foote, who was interested in the iron that shepherds in 1887 had found nearby. Foote had recognized the extraterrestrial origin of the metal and visited the site to see what else he could dig up. In addition to iron, he found microscopic diamonds. These had been formed on impact but Foote—not knowing this—had incorrectly thought the object that had hit had been as large as the Moon. Foote made a further mistake in that he didn't associate the crater with the meteorite material he was investigating. Despite his acceptance of the extraterrestrial origin of the material on the ground—in his mind, the nearby crater was a separate phenomenon that had been created by volcanic activity.

On the other hand, Gilbert, who had learned about the crater from Foote, was one of the first to propose that it had originated from a meteoroid. But in his attempt to scientifically verify his claim, Gilbert too came to the wrong conclusion. Since no one yet understood the morphology of impact craters, he incorrectly ruled out his impact hypothesis because the mass in the rim did not agree with the missing mass from the crater and also because the shape was circular and not elliptical as he would have predicted for an impact arriving from a particular direction. Furthermore, no one found any magnetic evidence for a difference in iron content that would have indicated something extraterrestrial. Given the lack of evidence of a meteoroid, Gilbert was compelled by his own methodology—which neglected the more subtle elements of impact crater formation that I will soon describe—to mistakenly conclude that volcanic activity and not an impact had been responsible for the crater.

The crater's origin was correctly pinned down at last when Barringer and Tilghman published a couple of extraordinary papers in the 1905 *Proceedings of the Academy of the Natural Sciences of Philadelphia,* in which they demonstrated that the Meteor Crater did indeed result from an extraterrestrial impact. Their evidence included the overturned rim strata, which I'm told is truly spectacular to see, and nickel oxide in the sediment. However, the thirty tons of oxidized iron meteorite fragments around the crater led Barringer to make a different and costly mistake. Barringer thought most of the rest of the iron would be buried underground and spent 27 years drilling to find it. Its discovery would have been another bonanza for the man who in 1894 had made fifteen million dollars (more than a billion in today's dollars) from the Commonwealth silver mine, which was also in Arizona.

But the meteorite was smaller than Barringer had thought and anyway most of a meteorite vaporizes on impact. So he didn't make any money or successfully convince many people of the crater's origin, even after his excavation was complete. Barringer died of a heart attack a few months after the president of the Meteor Crater Exploration and Mining Company—a venture Barringer had helped create—closed down operations. Barringer and his company lost $600,000 on prospecting the crater, but at least Barringer did live long enough to vindicate his hypothesis.

As planetary science advanced and people finally began to more fully understand crater formation, more scientists were won over to Barringer's deduction. The final confirmation came in 1960, when Eugene Merle Shoemaker—a key player in scientific understanding of impacts—found rare forms of silica at the crater that could only have arisen from rocks containing quartz that were severely shocked by impact pressure. Aside from a nuclear explosion—unlikely 50,000 years in the past—a meteoroid impact is the only possible known cause.

Indeed, Shoemaker carefully mapped the crater and showed the similarity between its geology and that produced around nuclear explosion craters in Nevada. His analysis legitimized the concept of an extraterrestrial impact on Earth and marked a milestone in Earth science's absorbing the significance of the Earth's interaction with its cosmic environment.

IMPACT CRATER FORMATION

My joy in rock climbing comes in no small part from my pleasure in examining the material, texture, and density of rocks—closely inspecting the surfaces to identify the safest and most efficient route up. But the real treasure buried in rocks is their long history. Along with the evidence of tectonic plate movement that they display, rocks' morphology and composition provide a trove of information for geologists to evaluate. Paleontologists too learn a great deal from the Earth's embedded fossils and its terrain.

Rock formations always tell a story, and some locations are particularly spectacular in this respect. On a recent university visit to the Basque country in Bilbao, Spain, I was very lucky to have a physics colleague tell me about the Flysch Geoparque in the nearby town of Zumaia. The Geoparque is an active ecotourism site that features an incredible limestone outcrop representing millions of years of geological history—fascinating both for its use of the geological treasures there to provide sustainable economic development and for its diverse scientific activity and discoveries. When I visited the Geoparque, the scientific director there pointed out to me the 60-million-year span of rock layers readily visible across the vertical cliff, which was wonderfully situated along a stunning beach. (See Figure 23.) He described the cliff as an open book with every page visible at the same time. The K-T boundary (now officially known as the K-Pg boundary, which I'll return to later) separates the layer of white rock with fossils from the grayer layer above without them. This line that marks the last major extinction is well preserved in this quiet site in the Basque country.

But such magnificent layers of rock aren't the only way to learn about the past. Impact craters, which are some of the most remarkable formations on the Earth's surface, constitute a very different wellspring of information. Despite our limited knowledge of how and when meteoroids will strike, scientists understand quite a lot about impact craters' geology. A crater's shape, rock morphology, and composition provide clues that help to distinguish impact craters from calderas or other round depressions. And, since impact craters' distinctive appearance and composition can be understood in large part from their origin, the depressions and special rock types where meteoroids made groundfall tell us a lot about the events that created the craters in the first place.

If the words hadn't already been corrupted by a spectacularly unsuccessful military policy, "shock and awe" would probably be the most cogent description of impact craters' formation. Impact

[FIGURE 23] The 60 million years of history visible in the rock of the Flysch Geopaque on Itzurun Beach in Zumaia, Spain. (Jon Urrestilla)

craters are the result of extraterrestrial objects hitting the Earth with sufficient energy to create a shock wave that excavates a circular crater—which is awesome indeed. The shock wave—not the direct impact—is responsible for the circular shape of impact craters. A more direct excavation would instead produce a depression with a preferred orientation that would reflect the impactor's initial direction—not something that looks the same all around. This was the red herring that misdirected Gilbert's Barringer Crater analysis. But the crater can't simply be understood as an impactor pushing down on rock. The crater is created when the impactor pushes down so much on the Earth that the compressed region acts like a piston, which rapidly decompresses to release stress, rebounding from the initial impact and ejecting material. The release of the pressure through the hemispheric pattern of the shock wave is the actual explosion that creates the crater. This subsurface explosion gives rise to the impact crater's distinctive circular form.

Objects that form impact craters usually hit the ground at speeds up to eight times the Earth's escape velocity, which is 11 km/sec, with roughly 20–25 km/sec most typical. For larger objects, this speed—many times the speed of sound—guarantees that an enormous amount of kinetic energy gets released, since kinetic energy grows not only with the mass but with the square of the speed. An impact on solid rock, which can be comparable to a nuclear blast, produces shock waves that compress both the object from space and the surface on Earth. The shock released on impact heats up the material it encounters and almost always melts and vaporizes the entering meteoroid, and—with sufficiently big meteoroids—regions of the target too.

The expanding supersonic wave creates stress levels far in excess of the local material's strength. This creates rare crystalline structures, such as shocked quartz, that are found only in impact craters—and in the blast region of nuclear explosions. (See Figure 24.) Other distinctive features include shatter cones in rock, which are conically shaped structures whose apexes point to the collision point, as illustrated in Figure 25. Shatter cones are also clear evidence of a high-pressure event that again can be accounted for only by impacts or nuclear events. Shatter cones are interesting in that they range from millimeters to meters in length, thereby providing a macroscopic scale effect in the material. Along with crystal deformations and evi-

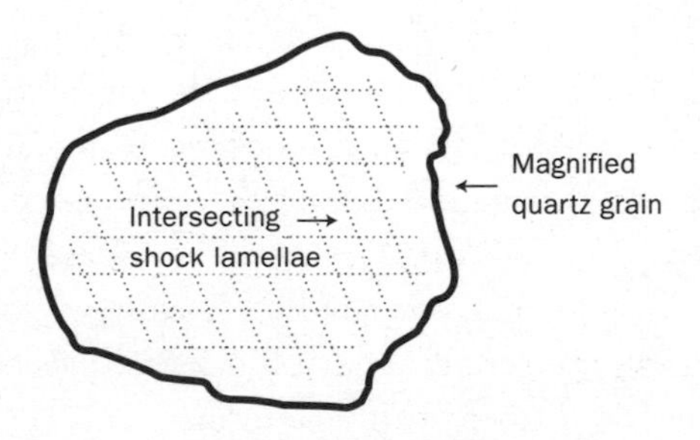

[FIGURE 24] The distinctive criss-cross deformation pattern in shocked quartz indicates a high-impact meteoroid origin.

dence of melted rock, shatter cones help distinguish craters that truly represent impact events.

Other rock forms that are characteristic of impacts are those formed at high temperatures. Known as *tektites* and *impact melt spherules,* these are glassy materials that have their origin in molten rock. As these are generated by high temperature and not necessarily high pressure, they can conceivably also originate in volcanoes—generally the chief contenders to impacts for crater formation. But impact craters generally have a different chemical composition, including metals and other materials—such as nickel, platinum, iridium, and cobalt—that are rare on the Earth's surface. These additional clues help corroborate an impact origin.

The chemical composition of the impactor can have other distinctive features as well. For example, particular isotopes—atoms with the same charge but different numbers of neutrons—might be more typical of extraterrestrial formations, though this will be useful only

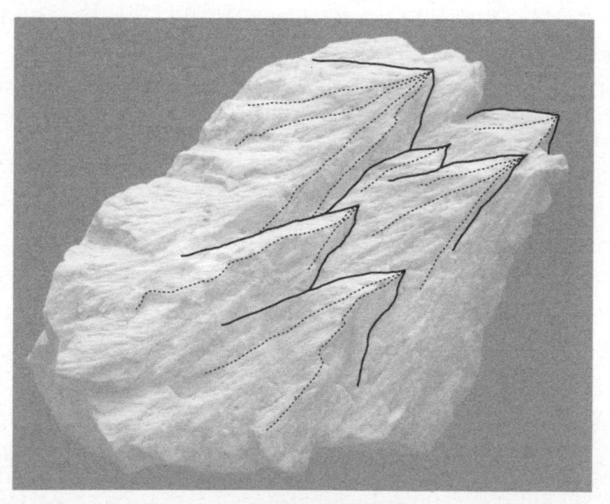

[FIGURE 25] The notable conical shapes of different sizes that occur many times in the same rock is a macroscopic indication of the rock structure's high-pressure formation.

for a small percentage of the remaining material since most of the original matter is vaporized

Also useful in distinguishing craters are impact *breccias,* which consist of fragments of rock held together by a fine-grained matrix of material—again indicating an impact that shattered what was initially there. Shocked fused glasses are also interesting in that their formation requires both high pressure and elevated temperature. Their unusually high density helps identify them. Another notable feature can be dikes within the crater floor or central sheets that line the floor of complex structures formed from glass particles.

These distinctive shock and melt features are critical to confirming impact events since there is no other way for them to form. However, they aren't always easy to find as they can be deeply buried under rock fragments and melt. Nonetheless, meteorites abound and many natural history museums display examples. I like the seven-foot-high, 34-ton Ahnighito meteorite at the American Museum of Natural History in New York, which is the largest on display. This enormous rock was a later acquisition, added to the meteorite collection that the museum has housed since its creation in 1869.

Materials help identify impact craters, as do their distinctive shapes. Whereas impact craters are depressions below the surrounding regions, most volcanic craters arise from eruptions and are found above the level of the surrounding terrain. Impact craters also have raised rims—again not typical for volcanic craters.

Another identifying feature is the *inverted stratigraphy*—the overturned rim strata—in the ejecta blanket, a consequence of material being excavated and then "flipping over" outside the crater—resembling the edge of a stack of large pancakes. The deep, roughly circular depression in the surface of the Earth—or on any planet or moon—with a raised rim and inverted stratigraphy is also clear evidence that a massive body hit the surface at enormous speed.

Though the materials distinguishing impact craters are mostly shaped during the sudden shock release, the crater's shape relies on the subsequent formation history too. Initially, when hitting the target, the impactor decelerates while the target material accelerates. The hit, compression, decompression, and the outflow of the shock wave all happen within a few tenths of a second. Once the shock wave has passed, changes occur more slowly. The accelerated material that got hit—having been accelerated by the initial wave—keeps moving even after the shock wave has decayed away, but at this stage the motion is subsonic. Even so, the crater continues to form, with the crater rim rising and more material ejected. The crater isn't yet stable, however, and gravity will make it collapse. For small craters, the rim falls down a bit and debris heads down the walls of the crater, with melted material flowing into the deeper portion of the crater. The end result is still bowl-shaped and looks a lot like the crater that initially formed, but it might be considerably smaller. The Meteor Crater, for example, is half its original size. Afterward, breccias and melted and ejected rock fill in the cavity. The shape of a simple crater is illustrated in Figure 26.

Larger impacts not only displace and eject material, but also vaporize part of the original ground that got hit. This melted material can coat the inside of the cavity, while the vaporized material typi-

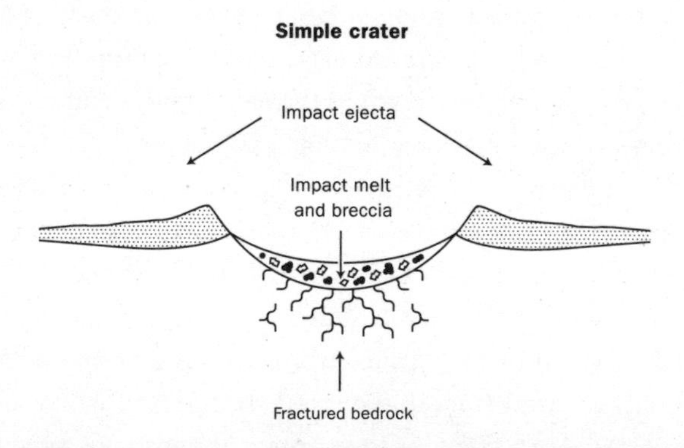

[FIGURE 26] A simple crater formed from an impact has an excavated bowl-shaped central region covered by relatively flat breccia and a distinctive raised rim.

cally expands away—creating in effect a mushroom cloud. Most of the coarser material will drop down within a few crater radii. But some of the more finely grained matter can be dispersed around the globe.

When the impactor is bigger than a kilometer across, the craters formed will be twenty kilometers or larger. The impactor in this case essentially creates a hole in the atmosphere, and ejecta fill this vacuum—going upward before descending over a wide area. The hottest material can rise above the stratosphere, and the fireball of vaporized material can then be widely dispersed, as happened to the worldwide iridium-rich clay deposited by the K-T impact we will encounter soon.

Larger impacts create a *complex crater* (see Figure 27), in which the cavity undergoes more extensive changes after the initial crater was established. The central region rises up while the rim partially collapses because as the shock wave propagates through the ground, it interacts with the nonuniform rock to generate a new wave that propagates in the opposite direction of the shock wave and "unloads" the shock. This rarefaction wave pulls deep material to shallow depth and leaves the crust thinned below large-impact craters. The speed at

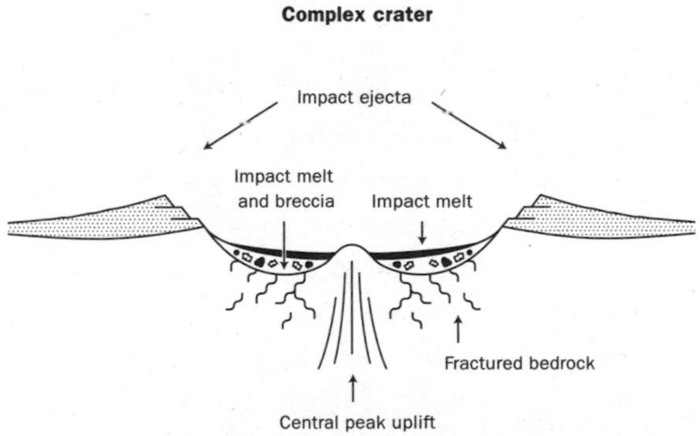

[FIGURE 27] A complex crater, like a simple crater, a raised rim—but with terraced structure—as well as an inner uplifted region and a greater quantity of collapsed material.

which all this happens is remarkable. Depressions several kilometers deep can be created in seconds, and peaks can rise thousands of meters in minutes.

Complex craters have an appearance different from the simple craters formed by smaller impacts. The precise crater shape depends on the size. When a crater is bigger than two kilometers across in layered sedimentary rock, or four kilometers across in stronger igneous or metamorphic crystalline rocks, it generally has a central uplifted region, a broad flat crater floor, and terraced walls. This is what is left after the initial compression, excavation, modification, and collapse.

When the size is larger than 12 kilometers across, an entire plateau or ring might rise up in the center. All these clues are critical information when (metaphorically and sometimes literally) digging up the past. As we will see in Chapter 12, these distinctive features were helpful in the 1980s in identifying the crater in the Yucatán associated with the K-T extinction.

CRATERS ON EARTH

Many impact craters have been found in the last half century. By studying their chemical composition as well as those of the scars known as *astroblemes*—the mostly destroyed craters that still leave identifiable impressions—we can begin to fill in our planet's visitor record. The guestbook is the Earth Impact Database.

The Earth Impact Database certainly contains some of the more fascinating lists you can find on the Internet. If you check it out, you will find catalogs of the many objects that have hit the Earth and left a scar big enough to be found and identified as an impact crater. This is not a complete list of impacts. Since a lot of the very old craters on Earth were wiped out by geological activity, most of the ones we observe here arose from more recent and less frequent impacts.

Most strikes probably occurred more than 3.9 billion years ago in the early phase of the Solar System when material leftover from

Crater Name	Age (millions of years)	Diameter (km)
Saint Martin	220 ± 32	40
Manicouagan	214 ± 1	85
Rochechouart	201 ± 2	23
Obolon'	169 ± 7	20
Puchezh-Katunki	167 ± 3	40
Morokweng	145.0 ± 0.8	70
Gosses Bluff	142.5 ± 0.8	22
Mjølnir	142.0 ± 2.6	40
Tunnunik (Prince Albert)	> 130, < 450	25
Tookoonoka	128 +/- 5	55
Carswell	115 ± 10	39
Steen River	91 ± 7	25
Lappajärvi	76.20 ± 0.29	23
Manson	74.1 ± 0.1	35
Kara	70.3 ± 2.2	65
Chicxulub	66 +/- 0.03	150
Boltysh	65.17 ± 0.64	24
Montagnais	50.50 ± 0.76	45
Kamensk	49.0 ± 0.2	25
Logancha	40 ± 20	20
Haughton	39	23
Mistastin	36.4 ± 4	28
Popigai	35.7 ± 0.2	90
Chesapeake Bay	35.3 ± 0.1	40
Ries	15.1 ± 0.1	24
Kara-Kul	< 5	52

[FIGURE 28] List of known craters on Earth greater than 20 kilometers in diameter created in the last 250 million years, obtained from the Earth impact data base. The sizes represent estimates for the diameter from rim-to-rim of the crater itself, which is smaller than the affected impact region.

the formation of the planets was swept up and moved around. But the Earth, Mars, Venus, and other more geologically active bodies have tended to lose evidence of the craters over time, which is why the geologically passive Moon shows them so much more prominently.

Even evidence of more recent impacts is mostly now lost. Though they occur reasonably often, small hits don't leave a noticeable scar—at least for long. In fact small craters are even less common than you might otherwise expect because of the Earth's dense atmosphere. Like on Venus and Titan, the atmosphere protects us from the many small impacts that occur more often on Mercury and the Moon, where the atmosphere does not protect them.

Larger impacts occur only rarely—pretty fortunate for the stability of life on the planet. An impact that is violent enough to produce a 20-kilometer-wide crater might occur and cause global damage once every few hundred thousand to a million years. Yet even this rate is not reflected in the Earth Impact Data Base. If you check it out, you will find evidence of only 43 such craters and only 34 within the last 500 million years, 26 within the last 250 million years (See Figure 28.), and only about 200 structures in total.

Several factors account for the paucity of the crater record. The first relevant issue is that 70 percent of the Earth's surface is covered by oceans. Not only are underwater craters difficult to find, but water can interfere with the formation of a crater in the first place. Furthermore, geological activity in the ocean floor is likely to wipe out all but the more recent of any scars that did in fact form. Seafloor evidence is largely eliminated every 200 million years, since plate tectonics changes the ocean floor in a conveyer-belt-like process of spreading and subduction that covers up any preexisting evidence on this time scale.

Even on the ground, geological activity such as erosion due to wind or water can destroy evidence. This is one reason most craters have been found in the more stable interior regions of continents (and are more likely to be retained on planets with less geological activity, such as Venus). And of course, even if not as inaccessible as four kilometers underwater, meteoroids might land in less accessible regions on land too. Finally, human processes can cover up evidence by changing the Earth's surface. So in some respects it's remarkable that the list of craters is as big as it is.

Several standouts are notable for being relatively recent events (on geological time scales). Two 10-kilometer-wide craters were created within the last million years—one in Kazakhstan and one in Ghana. Two other notables are those in South Africa and in Canada—Vredefort and Sudbury. These are even bigger than the Chicxulub crater caused by the impact event that gave rise to the K-T extinction, but were shaped in the far more distant past—a couple of billion years ago. The Sudbury mines in Canada were created to excavate the nickel and copper that became concentrated when the enormous object that created the crater struck and melted the crust. The impactor at Sudbury did not directly deliver most of the metals, but instead melted a huge sea-size volume of the Earth's crust that took a long time to crystallize. This left enough time for the small amounts of nickel and copper already present in the crust to settle to the bottom of the pool of impact melt. The metals were then further concentrated by hydrothermal activity generated by the hot impact melt sheet to produce economically recoverable ores.

The Sudbury mines are famous among particle physicists because of an underground lab there. Though still active mines, they are also an active physics experimental site. The Sudbury lab's deep underground location, two kilometers below the surface, shields detectors inside from cosmic rays, making the lab an ideal location to study neutrinos from the Sun, as it did between 1999 and 2006. It is also an excellent place to search for dark matter, which is the goal of several experiments currently housed inside.

But most impact stories are not quite so rosy. I'll soon describe the more recent Chicxulub impact event, which demonstrates the tremendous destructive capacity that large impacts can have. However, before presenting this incredible story of the meteoroid that triggered the K-T extinction 66 million years ago, let's first reflect on the larger tale of the major extinctions of the past half a billion years, and what they tell us about the fragility and stability of life on our planet.

11

EXTINCTIONS

Darwinian natural selection famously explains how life evolves. New species emerge while species that can't successfully compete and adapt to changes in their environment—or access an alternative suitable habitat—die out. Yet despite its many successes—and they are indeed numerous—Darwinian evolution does not completely account for life as we know it. The most critical missing element is life's origin.

Darwin helps us understand how some forms of life gave way to others once life had already emerged. But although evolutionary principles play a role, Darwin's ideas don't explain how life formed in the first place. Despite the many popular articles and books on the subject, origins are among the most difficult scientific questions to address—whether the issue is the start of life on Earth or the beginning of the Universe that contains it. Ideas about later stages of development are amenable to the scientific method in that they can be tested—if not always by controlled experiments in a laboratory, then at least by examinations of the fossil record or the rich and ancient sky. The beginning, on the other hand, is almost always inaccessible. Theoretically inclined scientists who interpret or more likely speculate about what came before might attempt to tackle origin questions. And some experimentally oriented biologists might try to replicate processes essential to the formation of life in the early Solar System. But despite this nascent progress, the beginnings of life remain—at least for the time being—very challenging to ascertain.

However, our focus in this chapter is on a different aspect of the story of life, one that—like life's origins—was not entirely encompassed by Darwin's initial theory of natural selection. This one, however, has the advantage that—like later-stage evolution—it is amenable to observations. This important element of the story of life concerns how it responds to radical change—including *mass extinctions,* in which many species die out at about the same time, leaving no direct descendants in their wake.

Central to the original Darwinian conception was the notion of *gradualism,* the idea that change occurs slowly over many generations. Darwin's theory wasn't about radical changes and it certainly didn't envision changes induced by extraterrestrial incursions. Darwin's picture relied on slow evolution, whereas environmental catastrophes can be very sudden. The current theory of evolution does allow for much more rapid change than Darwin originally envisioned. The Princeton biologists Peter and Rosemary Grant, following in Darwin's footsteps, famously found that the beaks of finches in the Galapagos Islands adapted quite quickly to changing precipitation—on a short enough time scale for the Grants to see the changes in their successive visits. But catastrophes generally occur so rapidly and with such dramatic implications that they make it impossible for many species to survive.

Dinosaurs had indeed adapted and, as a group, they survived for millions of years. Under other circumstances, they almost certainly would have survived many more. But they couldn't adapt to environmental conditions that they had never before experienced—which we will soon see originated in an object from outer space.

Studies of evolution now recognize that adaptation is almost always too slow a process to cope with any but the most gradual environmental changes. Adaptations seem to yield species with truly distinct properties only in isolated environments. The favored response to changing conditions is most often migration to a new location with a more suitable environment—but of course only if such an environment is accessible. When a species can't adapt or relocate to a suitable habitat, it doesn't stand a chance. In our rapidly changing environment, people would do well to take this into consideration. Technological advances notwithstanding, the lesson is probably relevant when evaluating the likely geopolitical implications of today's changing environment.

Closer to our cosmic tale, the story of extinctions is of interest to us because of the connections between life on our planet and our celestial, solar, and possibly galactic environment. It is easy to forget how dependent our existence is on the many contingencies that allow life to form—and to die out. This chapter discusses the notion of extinction, its causes, and the five largest mass extinction events in which a half to three-quarters of species died off (paleontologists haven't converged on a single definition) within a several-million-year time frame, as well as a sixth mass extinction that might well be in the works.

Extinctions connect our planet to meteorological events in both senses—weather and outer space. A better understanding of the connections is challenging, but could lie within our grasp. This science is important for us as a species—even if the stories unfold over longer time frames than most people tend to contemplate.

LIFE AND DEATH

Simple life emerged relatively early in the Earth's history. The oldest rocks on the planet's surface contain evidence of life in fossils that date from about 3.5 billion years ago—about a billion years after the Earth's formation and rather soon after it stopped being bombarded by asteroids and comets from space. Oxygenic photosynthesis emerged about a billion years later—and with it an atmosphere that most likely triggered many extinctions, but which also precipitated the emergence of multicellular algae. About half a billion years later began the "boring billion," in which no other radical development took place—at least that we know of. This long, tranquil interval suddenly ended at the beginning of the Cambrian period—about 540 million years ago—when complex life exploded.

Our more detailed understanding of evolution applies from the Cambrian diversification to the present day—in the time frame known as the *Phanerozoic eon*. The fossil record contains imprints right from its beginning, when many hard-shelled animals first appeared and generated a robust, lasting record, and most animal and plant life emerged. Fossils survive in regions as diverse as the Burgess Shale in the Canadian Rockies, the Yangtze Gorges of China, northeastern Siberia, and Namibia. All contain evidence for the widespread explosion of various types of life, as do the earlier Ediacara fossils in Australia, the Nama-type fossils in Namibia, the Avalon-type fossils in Newfoundland, and some fossils from the White Sea region of northwest Russia. These latter regions contain some of the earliest known complex life—from a time immediately preceding the Cambrian explosion.

The fossil record—in addition to telling us about life's proliferation—provides insights into times when different forms of life disappeared, without leaving any descendants. Although most of the fossils recording extinctions are very old, the notion of extinction is relatively new. Only in the early 1800s did the French naturalist and nobleman Georges Cuvier recognize the evidence that some species had entirely disappeared from the planet. Prior to Cuvier, when others had found older animal bones, they had invariably tried to relate them to existing species—which of course would have been a reasonable first guess. After all, mammoths and mastodons and elephants are different, but not so much so that you might not initially confuse them or at least try to relate their remains. Cuvier sorted it out when he demonstrated that mastodons and mammoths weren't ancestors of any currently living animals. He went on to identify many other now-extinct species as well.

But although the idea of extinctions is now firmly established, the idea that entire species could irreversibly disappear initially met with a lot of resistance. The concept of extinction must have been at least as difficult to reconcile with predominant beliefs back then as man-made climate change is for many today. The English geologist Charles Lyell, Charles Darwin, and Georges Cuvier all helped advance its acceptance—but not necessarily deliberately, and certainly from very different perspectives.

Cuvier, unlike the others, took the point of view that radical transitions in the fossil record were the consequences of Earth-wide catastrophes. Strong support for his viewpoint came from the observation that the rock at the point of rapid change in fossil types itself showed signs of cataclysmic events. Yet Cuvier didn't have the complete picture either. He over-zealously believed in his idea that all extinct species were done in by catastrophic events—never acknowledging that gradual changes could also be contributing factors. Cuvier refused to accept either Darwin's evolutionary theory or the idea that species frequently went extinct through slow, persistent processes.

In fairness, even now, people are confused when they see dramatic landscapes—not always appreciating the slow processes that helped shape them. A fellow speaker at an event in southwest Colorado remarked during the drive to the venue upon the dramatic upheavals that he imagined had created the vertiginous sandstone cliffs on either side of the road. I reminded him that the relevant processes occurred over millions of years—albeit in fits and starts—not nearly so drastically as he had pronounced.

At the time of Cuvier's proposal, most of the scientific establishment made the opposite mistake—opposing any role for catastrophic change. If the notion of extinction was a difficult pill to swallow a couple of centuries back, the idea of catastrophic change probably seemed more incredible still. Darwin was among those scientists who understood gradual change but omitted the very ideas that to Cuvier were so essential. Darwin assumed that any evidence contradicting gradualism was simply a sign of the inadequacy of the geological or fossil record. Though he of course accepted evolution, he assumed it always happened far too slowly for anyone to actively observe. Darwin in his thinking followed the point of view of the influential Charles Lyell, who in the second half of the nineteenth century was still arguing that all changes were smooth and gradual, reasoning that any so-called evidence to the contrary was simply imperfect data caused either by gaps in the geological record or by erosion. Lyell in turn was inspired in part by the Scottish physician, chemical manufacturer, agriculturalist, and geologist James Hutton, who thought the Earth changed only through tiny alterations that nevertheless yielded major effects over long periods of time.

These scientists' ideas are indeed correct for many processes—both biological and geological. Rain and wind slowly erode mountains, which themselves can be the result of a gradual uplift over millions of years that is precipitated by sluggish plate movement. But we now know that both gradual and rapid changes shape the planet, though even most radical changes would still rate as relatively slow from a human perspective. This is one of several reasons why these changes are so difficult to comprehend.

Yet with hindsight, we can look back and say the evidence for dramatic changes should have been obvious. Even as early as the 1840s, scientists had detected big gaps in the fossil record that were suggestive of catastrophic events. Paleontologists studying the sedimentary record identified such events when they noticed places where many fossil types suddenly ended at a boundary in the rock layer, above which evidence of new species began. This is not to say the evidence was always immediately unambiguous, since many phenomena might cause sedimentation to cease and then resume. But with the identification of corresponding cataclysmic events and careful dating that could determine the relative timing of the deposition of earlier and later layers, paleontologists could resolve many confusions. Over time, the evidence for rapid changes became too strong to refute.

OVERCOMING THE HURDLES

Scientists trying to reconstruct past events had to work extremely hard, however, to turn hypotheses into verifiable or falsifiable predictions. Even with an abundant fossil record, uncertainties in temporal or spatial resolution can point to very different hypotheses and conclusions. To understand the reasons for some of the continuing scientific debates—but also to appreciate just how clever and methodical the geologists and paleontologists who have surmounted these obstacles have been—let's briefly consider some of the challenges both in reliably ascertaining how quickly and extensively extinctions occurred and in determining the underlying cause.

The first obstacle is simply the difficulty in evaluating the rate of extinctions. Counting the precise number of species that exist on the planet at any given time is difficult since scientists would need to find, identify, and distinguish every type of mammal, reptile, fish, insect, and plant in existence. This applies even when counting existing species today, which should in principle be most accessible. The biologist E. O. Wilson in *The Future of Life* laments that there are too many discoveries of new species emerging every year for naturalists to write papers on all of them.

Somewhere between one and two million existing species have been catalogued, with the best estimate lying between eight and ten million—though estimates up to five times higher can also be found. Not surprisingly—given the gap in time and the problems with identifying not only life from years past but geological events and their influence—the extinction rate in the past is even more challenging to determine than the number of species existing today. After all, the number of past species and the rate at which they disappeared are then both harder to count.

A confusing technicality when identifying mass extinctions in particular is that the associated number can vary according to the precise definition. I'll mostly refer to the number of species, though scientists frequently prefer to count genera, which they might find the more useful grouping. My facility with understanding the relevant biological categories—important to both evolution and extinction—is greatly assisted by my long-ago preparation for a high school examination, when I memorized kingdom-phylum-class-order-family-genus-species simply by saying it enough times (try it). Despite my rarely calling on this knowledge, I have never forgotten these terms. These gradations, which might be unfamiliar to you, refer to how closely related are particular forms of life.

The category makes a difference when evaluating whether a mass extinction occurred. For example, consider a circumstance in which more than half the species in each genus are wiped out. Only one species in a given genus needs to remain for that genus to survive. If indeed this is what came to pass, species counting would dictate that an extinction event had occurred since more than half of the species were eliminated, whereas genus counting would indicate otherwise since that number didn't change at all. This example, along with the arbitrary percentage that is used to precisely demarcate a mass extinction event—some say 50 percent and some say 75 percent—illuminates the somewhat fuzzy nature of the definition. This is not to say mass extinctions can be ignored—just that there is no ideal way to define them.

The job of paleontologists is also burdened by more substantive factors than terminology. Clearly identifying and understanding disrupted fossil records is essential. If some species or genus leaves fossils in contiguous layers that are absent in the layers above, this would appear to signal an extinction event. But fossils are found only in sedimentary rock. The rare species that live in volcanic or other nonsedimentary environments generally won't leave any trace. An obstacle when studying older life-forms from before the Cambrian period (about 540 million years ago) is the absence of hard body parts, which makes identifying fossil deposits from earlier times very challenging.

But the more recent record is complicated too. Even if fossils do form, the interpretation can be confused by variations in the sedimentation and erosion rates that are essential to understanding fossils' implications. Terrestrial deposits occur episodically while erosion on land acts steadily, whereas in a marine environment, sedimentary deposits are constant and erosion is episodic. This makes the marine record more comprehensive than the terrestrial one, which is generally far less complete. These factors mean that only parts of the fossil record survive, and that record, even when present, can be difficult to find and identify. Paleontologists succeed only because even though the probability of finding any individual is very low, given enough individuals from sufficiently many species over a long enough period of time, the sedimentary record will still be replete with fossils.

Such fossils might be neatly preserved imprints of an entire individual, but more often they are only partial records—easily disguised evidence embedded in rock. Because generally only the hard parts of a species fossilize, the distinguishing body parts are frequently absent, causing different species to be conflated. Even if we were perfect at identifying fossils, erosion and other processes on Earth have readily hidden or destroyed many imprints well before they might have been found.

On top of this, the Signor-Lipps effect can confuse the interpretation. This phenomenon, named after Phil Signor and Jere Lipps, is tied to the rather intuitive idea that the last fossils of a species will be located at different geological times in different places, making the extinction tend to appear less abrupt and more gradual than it actually was. According to Signor and Lipps, the variation in depth of the last remnant fossils over a spatially extended region doesn't decisively establish whether an extinction occurred gradually or suddenly. This ambiguity can make the precipitating cause of a given extinction difficult to ascertain.

Researchers often prefer marine fossils because they are generally better preserved. In the nineteenth century, clams, ammonites, corals, and other fairly large species were most accessible, while in the twentieth century, geologists with more advanced tools started using microfossils like the one-celled *foraminifera*—which are abundant and widespread and preserved both underwater and in uplifted limestone—to get more detailed information.

A further consideration when establishing extinctions is that both the fossil records and the absolute ages are important. The fossil record in conjunction with the geological formations in which they are found can help establish relative ages. Since different time periods hosted different species, the types of fossils present help determine the relative times in which they were formed. But finding the absolute and not just relative ages of a boundary rock layer is often difficult and requires methods to date formations that are independent of the fossil record. One method that geologists often use for this purpose is *isotopic analysis*. With isotopic analysis, scientists determine the ratio of different isotopes of an atom (in which the number of protons is the same but the number of neutrons is different), reasoning that if you know how long it takes for one isotope to decay into another and you also know what you started with, you can then determine how old something is by how much of an atom type remains.

Carbon dating is perhaps the best-known example of this method. It is used to determine the age of older organic materials and is indeed very precise. However, given the half-life of carbon isotopes, it is effective only for stuff less than 50,000 years old, making it inadequate for dating the older rocks in most of the Phanerozoic. Longer-lived isotopes are used instead.

But isotopic analysis is more difficult when applied to older rocks. Generally only trace amounts of the relevant isotopes are present and the age determination is not always sufficiently precise. For example, potassium decaying to argon is an important process for dating. But argon gas in rocks can escape into the atmosphere, making rocks appear younger than they actually are. Or it can be trapped in the rock when first formed, leading to more argon and the appearance of an older formation. Methods have significantly improved in the last few decades as cross-correlations of various elements have made studies even better and detailed probes of even minute elements becomes more accessible. The recent dating of the meteoroid and the K-T extinction event using lasers to remove gas from argon crystals, which I'll mention in the following chapter, provides a spectacularly accurate example.

Magnetic information has also been used to help establish absolute ages. This method, which was initially employed for rocks relevant to the dinosaur extinction, relies on geomagnetic reversals. But because the Earth's crust is made of moving tectonic plates, the magnetic field's orientation changes over time so that the initial orientation can be hard to reconstruct, compromising the reliability of the results. Perhaps that was a good thing in that its inadequacy precipitated the search for another method by the geologist Walter Alvarez and his father, the physicist Luis Alvarez, which led to the meteoroid hypothesis that I will soon explain.

PROPOSED EXPLANATIONS FOR EXTINCTIONS

The hard work of geologists and paleontologists has unquestionably demonstrated that spectacular changes occurred in the past that wiped out most life on the planet. Once this had been established, the questions turned to how and why they occurred. We've experienced a number of devastating storms and disasters in recent years, but none individually has been nearly powerful enough to eliminate half the species on the planet. Of course the final verdict on the cumulative effect of human influence is yet to be determined. But what precipitated the world-changing disasters that have occurred in the past?

Before getting to a list of cataclysmic events that can trigger extinctions, let's first consider the rather short list of environmental factors that might come into play. Temperature or precipitation changes (in either direction) have been two important contributors. Broadly speaking, species that have adapted to their local environment are not necessarily able to adapt when weather patterns change.

As with the melting of the arctic ice, the environment for particular species can change so dramatically in response to the changing temperature that existing species, which can't adapt sufficiently quickly, have to move to a suitable habitat—or perish. Climate change has less direct effects too of course—one of the most significant being the change in sea levels, which can destroy stable marine environments and flood previously habitable landmasses—turning terrestrial environments into oceanic ones and thereby eliminating some land-dwelling species.

Warming oceans can affect precipitation patterns too—again influencing species' chances of survival. On shorter time scales, parasites or diseases—the dangers of which can be exacerbated by climate change—can also contribute to extinctions. And the food on which a species depends might die off, triggering a domino effect in the food chain.

In the oceans, change in acidity is a further potential kill mechanism, as is oxygen depletion. Finally the formation of barriers that can lead to isolated, vulnerable populations or the removal of barriers that can permit invasive species or overhomogenizing of populations are two other factors that can doom a species. Any extinction trigger causes at least one of the disasters I've just described, and most of them cause several in combination.

But why do these changes occur? What environmental change triggered them? Two competing viewpoints dominate the thinking on the subject. One is that they occurred gradually—a point of view often connected to Earth-bound phenomena such as volcanoes or plate tectonics. The dust and soot that a volcano emits can block sunlight and produce significant enough changes in the atmosphere to affect the temperature. But it can take a while for life-forms to die out as a consequence. Plate tectonics, which can influence habitats and environments, is another suggested cause of the gradual eradication of species. Along with changes in oceans, plate tectonics can affect climate and land coverage, and both of these can lead to dramatic modifications to life on the planet. Of course if either volcanism or plate tectonics is relevant to extinctions, most likely both are, since they tend to happen concurrently.

Then there are the "big events." This opposing viewpoint encompasses externally imposed extraterrestrial catastrophes such as large meteoroid impacts, but it can include destructive terrestrial events too if they happen sufficiently suddenly. Proposals for terrestrially induced catastrophes rely on well-known phenomena that might suddenly transpire at an accelerated pace. For example, we know volcanoes erupt at various intervals, but in Siberia and on the Deccan Plateau of southern India, basaltic lava layers extend over an enormous region called a *trap*. A trap contains lava layers that extend into a large plateau, which arose from extremely high volcano eruption rates that emitted enormous amounts of lava that spread over a large region. The Deccan and Siberian traps are signs of much more violent volcanic activity and eruption rates than usual. Even today, despite erosion, the lava from the Siberian traps occupies a million square kilometers at least and a few hundred thousand cubic kilometers volume.

Ash from the kind of extensive volcanic activity that formed the traps would have caused serious damage. You might recall from news reports how ash can spread densely enough to interfere with aircraft, as happened in April 2010 when the Icelandic volcano Eyjafjallajökull erupted. More vigorous volcanic activity can give rise to more substantial, global effects, such as significant changes in the planet's weather. Eruptions emit large amounts of sulfur dioxide. This might increase the amount of water vapor in the upper atmosphere and contribute to a greenhouse effect and hence global warming on short time scales. On longer time scales, these same volcanoes might cause global cooling. That's because the sulfur dioxide that is emitted rapidly combines with water to create sulfuric acid. This condenses to form fine sulfate aerosols that radiate sunlight back into space, thereby cooling the lower atmosphere. (It works so well that deliberately injecting sulfur into the atmosphere is one strategy that scientists are investigating as a precarious climate engineering response to climate change.) Sulfate aerosols can also destroy the ozone in the atmosphere and lead to acid rain. Feedback effects—both known and unknown—might also create even more long-lasting weather phenomena.

However, volcanic activity alone won't explain all extinctions. Activity extensive enough to destroy the majority of life on Earth is rare. More exotic suggestions for triggers of rapid and catastrophic mass extinctions center on cosmic events. Changes in the Earth's axis and orbit occur too and are responsible for some of the climate changes, such as ice ages, that occur on time scales of tens or hundreds of thousands of years, but these Earth motions are unlikely to explain the massively destructive events that occur so much less frequently.

Cosmic rays and supernovae as well as cosmic impacts have been suggested as possible culprits with the potential to be relevant on longer time scales. Cosmic rays can affect cloud cover in several ways. One is by ionizing atoms in the troposphere so that water droplets can nucleate nearby. This influence could enhance cloud formation, which would in turn affect the Earth's weather. However, this theory doesn't necessarily hold water (so to speak). First of all, we don't know how important cosmic rays are relative to other potential ionizing sources. Second, the nuclei—even when formed—have to grow enormously by condensation before they are truly forming clouds. Third, the effect of clouds isn't clear—they could cool the Earth by reflecting sunlight or, alternatively, they could heat it further by reradiating some of the energy. In any case, correlations that have been measured between cosmic rays and climate don't suffice to explain the enormous changes over a short time period required to precipitate extinctions.

Supernovae have also been suggested as possible extraterrestrial extinction triggers. The proposed mechanism concerns the energetic X-rays and cosmic rays that supernovae release. This radiation could in principle kill life directly by destroying cellular or genetic material. The radiation could also deplete the ozone layer or lead to nitrogen dioxide formation, which would in turn lead to global cooling by absorbing sunlight.

However, despite these potential dangers, supernovae are unlikely to explain extinctions, for exactly the reason you might suspect: supernovae that are close enough to cause major problems don't occur sufficiently frequently. Although the supernova rate increases when the Earth passes through the spiral arms of the galaxy where the density of supernovae is greater, the likelihood of the Earth passing close enough to them is still too low to account for extinction events. Similarly, gamma ray bursts don't occur frequently enough to explain most extinctions. According to some estimates, they occur only every billion years or so in the Milky Way.

A far more compelling candidate for a cosmic extinction trigger is a comet or an asteroid that makes impact on Earth. An enormous object hitting the planet can precipitate dramatic changes on the ground, in the air, and in the oceans. If something sufficiently big hits, major—and for some species fatal—changes in the Earth's surface and climate would follow immediately.

In fact, essentially most of the great movie disaster scenarios (with the exception of a zombie apocalypse) follow in the wake of a sufficiently big impact. The impact itself creates shock blasts, fires, earthquakes, and tsunamis. Dust can block out the atmosphere, temporarily ending photosynthesis and eliminating the majority of the food sources for most animals. Climate changes induced by the strike wreak havoc as well—both initial heating and later cooling and possible further heating after that. Cooling comes from sulfates and dust that remain in the atmosphere, while the later heating is a possible result of toxic and heat-trapping gases that could set off global warming.

A meteoroid did indeed cause at least one mass extinction event that the next chapter will explain in detail. This calamity was one of the five big extinction events that took place during the Phanerozoic eon.

THE BIG FIVE

In 1982, the University of Chicago paleontologists Jack Sepkoski and David M. Raup revolutionized the field of paleobiology with their pioneering quantitative analysis of all the field's existing data. The many imperfections in the observations—and the many decisions that are needed to decide what to include and how—made their numerical, data-driven study tricky. Yet they realized that statistics can be useful even when applied to imperfect or incomplete data if enough of it is available, and that was indeed the case. Although the 1982 Raup and Sepkoski paper was not the first quantitative study of the fossil record, it reoriented the study of extinctions, which had formerly relied far more on narrative and small-scale investigations.

In their research, the Chicago paleontologists identified five major mass extinctions (see Figure 29) and about twenty lesser ones, in which approximately 20 percent of life-forms died out. Because of the very strong differences in evolutionary dynamics and the less available and reliable evidence from earlier on, Raup and Sepkoski focused on life—and its destruction—in the past 540 million years. The emergence and disappearance of life-forms certainly occurred before the Cambrian explosion too. But the spotty fossil record makes counting the species from very early times a nonstarter.

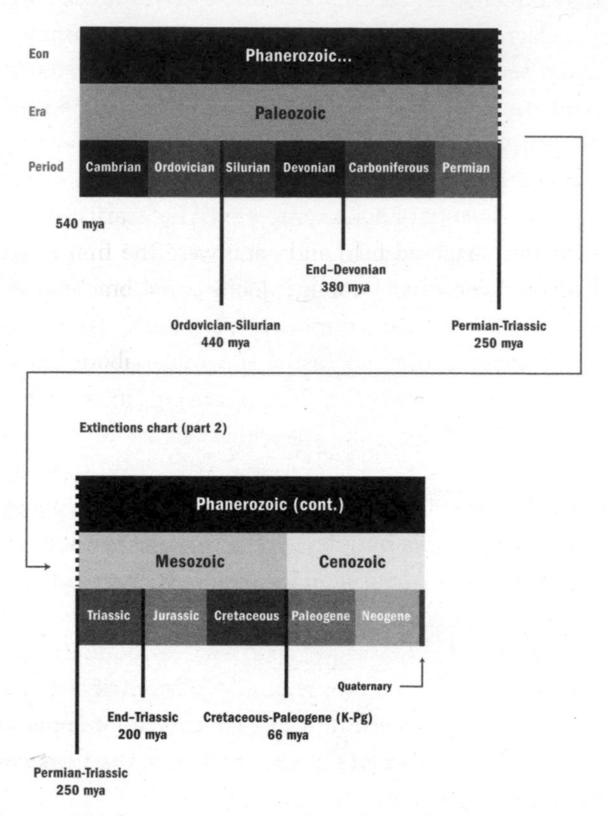

[FIGURE 29] The five major extinction boundaries: the Ordovician-Silurian about 440 million years ago, the late Devonian about 380 million years ago, the Permian-Triassic about 250 million years ago, the end Triassic about 200 million years ago, and the K-Pg extinction about 66 million years ago. Also shown are the eras and periods of the Phanerozoic eon.

The oldest major event they identified is the Ordovician-Silurian extinction, which occurred somewhere between 450 and 400 million years ago. Essentially all life was in the ocean back then so most of the species lost were marine-based. This mass extinction—the second most deadly, in which about 85 percent of all species went extinct—occurred in two stages over a period of about 3.5 million years. The cause seems to have initially been lower temperatures and massive glaciations that caused dramatic drops in sea level. Such drops take place when water becomes trapped as ice—the opposite of the feared sea level rise that will occur in the near future when glaciers melt and convert ice to water. The second extinction pulse was probably due to a later warming period that wiped out the fauna that had adapted to the cold. Warm-adapted fauna such as tropical plankton, shallow-water crinoids (predecessors of starfish and sea urchins), trilobites, armored fish, and coral were the first to go, then the cold-adapted versions of coral, trilobites, and brachiopods went after.

The next mass extinction lasted a while—about 20 million years—and started about 380 million years ago in the late Devonian period near the Devonian-Carboniferous transition. It seems there were a number—the number is uncertain, but suggestions vary from three to seven—of pulsed extinctions, each lasting about a few million years. This event hit marine life hard too, killing a significant fraction of species that lived in the oceans. Insects, plants, and early proto-amphibians survived on land, though extinctions were rampant there as well. Paleontologists think that one distinguishing aspect of this mass extinction is that it was primarily the result of a significantly smaller speciation rate that couldn't compensate for the steady expected rate of loss of species, which wasn't necessarily much higher than usual.

The Permian-Triassic (P-Tr) event about 250 million years ago was the most devastating known extinction in terms of the percentage of species that disappeared from the planet. Life, including amphibians and reptiles, had blossomed both in the sea and on land for quite some time after the Devonian extinction. But that ended during the P-Tr extinction, when at least 90 percent and probably more of the species on both land and in the sea died off. The losses included surface plankton as well as bottom-dwelling species such as fixed bryozoans and corals, some shellfish, and trilobites—species that had already survived two major extinction events. On land, even insects were devastated—the only time they suffered a mass extinction. In addition, a large portion of amphibians disappeared, and reptiles—who had only emerged after the previous extinction—lost most of their member species too.

The cause of the P-Tr extinction remains the subject of controversy, but massive climate change and changes in the chemistry of the atmosphere and oceans almost certainly played a role. Though the cause and mechanism are unclear, the approximately eight degree Celsius rise in temperature was likely due at least in part to massive volcanism in Siberia and the consequent enormous amount of carbon dioxide and methane emissions from the Siberian Traps. The Permian-Triassic extinction—the biggest known extinction event—is almost certainly partly the result of these volcano-induced gases that heated the planet, stressed the oceans, reduced oxygen, and poisoned the atmosphere. Even today, after what must have been a lot of erosion, the lava from the Siberian traps occupies a million square kilometers at least and a few hundred thousand cubic kilometers volume. At that time the traps were more like the size of present-day Russia.

Though existing life was nearly wiped out, what's bad for the goose can be good for the gander. Ferns and fungi replaced earlier flora and new plants eventually emerged. Mammal-like reptiles were no longer dominant after this period, but the modern mammal group developed from them. The emergence of archosaurs was another very significant consequence, eventually leading to the dominance of the dinosaurs.

A friend recently proudly showed me an extremely well-preserved—and kind of adorable—six-inch-long fossil that she told me was a 300-million-year-old dinosaur. Had she shown it to me a year earlier, I would have simply admired the details. But in light of my recent research I knew that the description couldn't possibly be correct since dinosaurs emerged only in the Triassic period, more recently than 250 million years ago. Enamored with the idea that it was indeed a dinosaur, I suggested the fossil was perhaps a bit younger. But it turned out the mistake was different: the fossil was indeed 300 million years old, but it wasn't a dinosaur—it was a mesosaurus, an extinct reptile species. Dinosaur fossils are old—but not so old as her well-preserved rock imprint.

Given the magnitude of the P-Tr devastation, life on Earth didn't recover very rapidly. We know this because the black shale, which extends several meters above the boundary sedimentary layer that demarcates the extinction, demonstrates a prolonged absence of white limestone-producing life forms. Nonetheless, after at least five million years, new forms of mollusks, fish, insects, plants, amphibians, reptiles, early mammals, and dinosaurs emerged. But this blossoming of life was interrupted within 40 or 50 million years by the fourth mass extinction, which arrived about 200 million years ago.

Roughly 75 percent of all species went extinct in this end-Triassic extinction that preceded the Jurassic period. The cause of this extinction is uncertain, but lower sea levels and the start of the volcanic rift that eventually produced the Atlantic Ocean might have played a role. Most large vertebrate predators in the ocean were killed off and species of sponges, chorals, brachiopods, nautiloids, and ammonoids were badly hit as well. This extinction also eliminated most mammal-like creatures, many large amphibians, and nondinosaur archosaurs.

The removal of any real competition on land left the dinosaurs essentially in charge. Extinctions destroy life, but they also reset the conditions for life's evolution. The Jurassic period that followed is famous for the book and movies named for it, even if not all the creatures featured in *Jurassic Park* lived during that interval. But the Jurassic period is indeed the time when dinosaurs rose to prominence—achieving dominance over the ecosystem on land by the late Jurassic. Flying reptiles, crocodiles, turtles, and lizards multiplied in this period and mammals evolved too, though their days in the limelight would have to await the mass extinction that would follow.

The most recent mass extinction is probably also the most famous. It is the one that occurred at the boundary between the Cretaceous and Paleogene periods. This event, formerly known as the K-T extinction (which stood for Cretaceous-Tertiary) but now officially known as the K-Pg extinction (because the Tertiary Period was renamed the Paleogene), occurred 66 million years ago. It is most commonly known as the one that killed off the dinosaurs.

But dinosaurs weren't the only species to go extinct. About three-quarters of the species and half the genera alive at the time disappeared, including many reptiles, mammals, plants, and marine life. Microscopic marine fossils, the most common in the sedimentary record, are especially important since their abundances can yield a detailed record of what took place. Each centimeter of marine deposits can be resolved to about 10,000 years of activity, giving a fine-grained picture of what occurred in the oceans. The internationally organized Ocean Drilling Program examines cores with resolution ten times as accurate as that. The accurate microfossil scale helped determine that plankton, coral, bony fish, ammonites, most sea turtles, and many crocodiles disappeared then as well.

Following the K-Pg extinction, mammals became a much bigger player on Earth. Though many factors contributed, the elimination of the terrestrial dinosaurs was almost certainly relevant. Large mammals (such as ourselves) might never have risen to prominence had terrestrial dinosaurs who dominated the competition for essential resources not first been eliminated. One speculation for why dinosaurs fared so much better than mammals before the Chicxulub impact is that dinosaurs laid eggs in volume, whereas mammals have fewer offspring and give birth less frequently the larger they are. Dinosaurs might have simply won a numbers game in their competition with large mammals.

Because it is the most recent—and because it is the one that let large mammals take over—the K-Pg extinction is the most carefully studied of the five known mass extinction events. The search for the correct theory to explain this global disappearance of both land and marine life is a great story that we'll turn to in the following chapter. The almost certain answer is that a huge meteoroid hit the planet 66 million years ago. Though clearly a part of the distant past, the separation in time is about as long as a year is to a 50-year old when compared to the four-billion-year life span of the Earth. I find it remarkable that this extraterrestrial intervention with such enormous consequences affected the Earth so (relatively) recently.

A SIXTH EXTINCTION?

But disaster might be even closer at hand. I would be remiss in my moral obligations if I were to close this chapter without presenting one final and very disturbing speculation. Many scientists today think we are currently undergoing a sixth mass extinction—of man-made origin. To definitely establish this claim would require determining the number of species that currently exist and the rate at which they are disappearing—both of which are difficult, if not impossible tasks. Yet even if inconclusive, the numbers we do have certainly indicate disturbing trends. The evidence points to a significantly higher extinction rate than usual, with current rates of species loss in line with those measured during past extinctions. At the estimated baseline rate, we would expect about one extinction a year. Estimates are uncertain, but the rate today might be hundreds of times greater than average.

If the measured rates of avian, amphibian, and mammalian extinctions are representative of what is to come, they are very disturbing. Mammals comprise a small fraction of the total number of species, but they are the ones that are best measured. Within the last 500 years, 80 mammal species have gone extinct out of a total of less than 6,000.

This mammal extinction rate of the last 500 years has been about 16 times higher than normal, and in the last century the rate has been elevated by a factor of 32. Amphibians in the last century have died off at a rate nearly 100 times higher than in the past, with 41 percent currently facing the threat of extinction, while bird extinctions in this same time frame have exceeded the average rate by a factor of about 20.

These numbers are consistent with an extinction event. As the University of California, Berkeley biologist Anthony Barnosky and others have observed, so too are the changes to the environment that are occurring now, which have a disturbing resemblance to those at the time of the P-Tr extinction. Carbon dioxide levels rose at that time—as did the temperature—oceans became more acidic, and dead zones where oxygen is absent arose in marine environments. Incredibly, the rate of temperature and changes in pH (which measures acidity) seem to have been comparable at that time to what they are today.

Human influence is almost certainly largely to blame for the recent diversity loss. People affect the planet and its life-forms in many ways. When Europeans arrived in North America, for instance, 80 percent of its large animals went extinct—in large part due to outright killings. But humans damage habitats in other ways too. One culprit is pollution, another is land clearing—including deforestation and overfishing, and another factor is climate change through both changes in temperature and changes in sea levels. Droughts and fires, floods and storms, as well as warmer and more acidic oceans all are relevant to which species survive. The human destruction of habitats facilitates species invasions on a local level and homogenizes populations on a global one—making any disease or parasite far more dangerous. Species move to new habitats when they can, but if those habitats are destroyed, so too will be the potential inhabitants. Given all these detrimental impacts, an imminent crisis in populations is certainly not out of the question.

Barnosky makes the interesting argument that the staggering human population growth that precipitated the current population crisis correlates in an interesting way to our energy consumption. Assuming an equitable distribution of resources and a reasonable estimate of the size and range of large mammals, the energy coming from the Sun each day supports a given number of animals and species. The number of *megafauna* species began to drop from about 350 to half that number between 50,000 and 10,000 years ago when humans showed up on the planet and usurped a disproportionate fraction of its resources. Subsequently, the number of mammals slowly rose back to its former value, but then began to escalate rapidly about 300 years ago—essentially when the industrial revolution allowed humans to dig into our energy savings—the unused energy that has been stored over millions of years in fossil fuels, the name of which is not incidental. And with that dip into our reserves—though the number of species has declined—the human and livestock populations have exploded along with urbanization.

Some optimists—while acknowledging the disturbing trends—argue that we might also create or resurrect species by designing or reproducing DNA and avoid a mass extinction (defined by the fractional change in species or genus number) by compensating with new species for the species that are lost. But revival of what was actually there before will be very challenging, given how poorly DNA is preserved and how unlikely re-creating the environment any past species lived in would be. Moreover, the rate at which we could create new species that could survive is unlikely to rival the pace at which the world is losing them. In any case, extinction is just a word. The assessment, which is based on one number, fails to capture the huge change that this (admittedly rather unlikely) scenario would entail.

Another way to avoid extinction, technically speaking, is for the trend to reverse before the numbers are depleted by half. For example, once the numbers have thinned out enough, maybe species will survive that wouldn't have been able to compete in a more biodiverse environment. This "optimistic" scenario is just a speculation in any case and, furthermore, the only saving factor would be the significant loss of life that would precede an ultimately more stable environment.

Such changes might ultimately be good for some future species. After all, even the P-Tr extinction left some life intact. From a dinosaur's perspective, for example, it was a very good thing. However, that doesn't obviate the loss of life that precipitated it, or the fallow interim time of suffering and mayhem during which life recovered. Though the consequences of the changes we are currently inducing could likewise ultimately be beneficial in some global sense, they won't necessarily be so for the species on Earth who have developed to adapt to the way things are now.

Even if new species do emerge or conditions ultimately improve, a dramatically altered world is unlikely to be good for us as a species. It does seem misguided for humans to be responsible for so much loss of biodiversity, which—through the consequent loss of food, medicine, and clean air and water—is likely to hurt ourselves. Life has evolved with delicate balancing mechanisms. It is not clear how many of these can be altered without dramatically changing the ecosystem and life on the planet. You would think we would have considerably more selfish concern for our fate—especially when so many such losses can most likely be prevented. After all, unlike the creatures 66 million years ago whose fate was determined by an errant meteoroid, humans today should have the capacity to see what is coming.

12

THE END OF THE DINOSAURS

Everyone loves the dinosaur. Whether they appear in skeletal, in fossil, or even in molded-plastic form, both young and old are captivated. Kids love these creatures from the past—building models and memorizing names that most adults can barely pronounce. A museum anywhere with a dinosaur display will exhibit heavy foot traffic—from both kids and their more senior counterparts. Curators at natural history museums are well aware of the attraction of these bizarre ancient reptiles. The American Museum of Natural History in New York counts among its chief attractions the enormous skeletons of a *Tyrannosaurus rex* (which translates as "king of the lizards") and an apatosaurus, as well as related models that greet you in the entryway.

Further evidence of their popularity is the starring role that dinosaurs play in popular culture—from Dino in *The Flintstones* (no, terrestrial dinosaurs didn't coexist with people) to the regenerated dinosaurs of *Jurassic Park* (no, they probably won't in the future either). Even the filmmakers of *King Kong* weren't satisfied with a giant ape that could scale the Empire State Building. They had to include a completely superfluous (opinions expressed represent those of the author only) scene with dinosaurs.

Why? Because dinosaurs were amazing. They looked enough like current animals to seem familiar, but were different enough to be exotic and weird in ways that activate our imaginations. They had horns and crests and bony armor and spines. Some were big and slow and others were small and swift. Some walked on the ground—some on two legs and some on four—while some of them flew in the air.

Yet for many people, the first thing that comes to mind when they think about dinosaurs is that those magnificent animals no longer walk the face of the Earth. Though dinosaurs did evolve into birds that survive today, the dinosaurs that had dominated the land for millions of years went extinct about 66 million years ago. Some people even regard the dinosaurs' demise with a touch of superiority—how could it be that such strong and agile creatures were foolish enough to disappear? Yet the fact is the dinosaurs were the major players on the planet for far longer than humans or apes are likely to survive. When they disappeared, it was through no fault of their own.

The question of what caused the land-dwelling dinosaurs to depart the planet was for a long time a tremendous mystery that mesmerized both scientists and the public alike. Why would this diverse, vigorous group that seemed to have mastered its environment suddenly disappear at the end of the Cretaceous period? This topic might seem a distant remove from physics—especially that associated with dark matter. But this chapter presents the many pieces of evidence that have demonstrated that a meteoroid impact was almost certainly the culprit—connecting this extinction to an extraterrestrial object in the Solar System. And, if the more speculative work that I did with my collaborators turns out to be correct, a disk of dark matter in the plane of the Milky Way was responsible for triggering the meteoroid's fatal trajectory. Whatever the role that dark matter had, the impact of an object from outer space that wiped out at least half the species on the planet certainly occurred—binding this extinction to our solar environment. The tale of how geologists, physicists, chemists, and paleontologists came to this conclusion is one of the best stories in modern science.

DINOSAUR TIME

The dinosaurs—apart from their range of sizes and their coolness—were striking as a category in their longevity, dominating the planet for more than 100 million years. Yet despite their apparent robustness and the proliferation of flora and fauna that accompanied them, a great deal of life ended abruptly 66 million years ago. The questions that persisted until late in the twentieth century were why this happened, and how.

Before answering these questions, let's first reflect on the age of the dinosaurs, and how different the Earth was back then. Dinosaurs lived in the Mesozoic era, which ranged from 252 to 66 million years ago. (See Figure 29.) The name *Mesozoic* comes from the Greek term for "middle life" and indeed this era lies in the middle of the three geological eras of the Phanerozoic eon. The Mesozoic era is wedged between the *Paleozoic,* with its name meaning "ancient life," and the *Cenozoic* whose name refers to "new life." This bracketing reflects the most devastating mass extinction we know of—the Permian-Triassic extinction event, which defines the first boundary, and the Cretaceous-Paleogene extinction—formerly known as the K-T extinction—which defines the second one, in which the (non-avian) dinosaurs and many other species disappeared.

The *K* in *K-T* comes from the German word *Kreide,* which means chalk. Similarly, the word "Cretaceous" to which it refers comes from the Latin word *creta,* which is literally Cretan earth—also meaning chalk. The *T* on the other hand comes from *Tertiary*—a relic from a now defunct naming scheme that divided the Earth's history into four parts, of which the Tertiary was the third.* Even so, like

---

* The International Commission on Stratigraphy (ICS), which is responsible for naming these time frames, also tried to eliminate the fourth subdivision—the Quaternary—but the International Union for Quaternary Research protested. So in 2009 the ICS restored the term. The Tertiary period—with its less ardent defenders—is no longer an official term, which is why *K-T* was replaced with *K-Pg.*

many others, I occasionally fall back on the colloquial term *K-T* for the extinction, though I will usually use the more correct term—K-Pg—from now on.

Eras are divided into periods that are further divided into epochs and stages. The Mesozoic era is subdivided into three periods: the Triassic period—lasting from 252 to 201 million years ago, the Jurassic period—from about 201 to 145 million years ago, and the Cretaceous period—from 145 to 66 million years past. "Mesozoic Park" might even have been a more accurate name for the Michael Crichton–Steven Spielberg movie, which features two Jurassic dinosaurs but also several that did not emerge until the Cretaceous period. I will nonetheless concede that "Jurassic Park" has a better ring to it, so I won't question the wisdom of the choice.

A lot changed on Earth during the Mesozoic era. Warming and cooling as well as significant tectonic activity transformed the atmosphere and the shape of the landmasses. The supercontinent Pangaea split in the Mesozoic era into the continents we see today, and resulted in extensive land movement over time.

Even though tectonic movement in the late Cretaceous brought the planet closer to its current state, the continents and oceans were not yet in their current positions. India had yet to collide with Asia and the Atlantic Ocean was much narrower. As tectonic plates have moved since then, oceans have changed size at the rate of several centimeters each year.

This effect alone tells us that 66 million years ago, most shores were several thousand kilometers away from their current locations—so, for example, the Americas and Europe were much closer. Moreover, sea levels were probably a hundred meters higher than they are today. Temperatures—especially in regions far from the ocean—were higher then as well. These factors turned out to be critical to deciphering some of the clues revealed at the K-Pg boundary. Although we now know that at the time of its formation, Italian sediment containing the clay that the Berkeley geologist Walter Alvarez had decided to study was part of a continental shelf that lay under hundreds of meters of water, researchers initially didn't know this to be the case.

Life on Earth evolved in response to its changing environment. The many moving pieces of land separated by water allowed new species to emerge. During the Triassic period, arthropods, turtles, crocodilians, lizards, bony fishes, sea urchins, marine reptiles, and the first mammal-like reptiles arose. The Late Triassic is also when many distinct species of dinosaurs, including terrestrial dinosaurs, first appeared. They went on to become the dominant land vertebrates during the Jurassic period.

Birds too emerged during this time, evolving from a branch of theropod dinosaurs. *Jurassic Park* doesn't necessarily get all the science right, but the film taught many that birds evolved from dinosaurs. Flying reptiles, marine reptiles, amphibians, lizards, crocodilians, and dinosaurs continued into the Cretaceous period, during which time snakes and early birds first appeared, as did flying reptiles and gingkoes, as well as modern plants such as cycads, conifers, redwoods, cypresses, and yews, whose forms we still observe today. Mammals also made an appearance, but they were small—generally between the size of a cat and a mouse. This changed only after the dinosaurs went extinct and left room and resources for them to evolve to bigger-bodied animals.

LOOKING FOR ANSWERS

Two fascinating books I read while working on this one were the geologist Walter Alvarez's *T. rex and the Crater of Doom* and the science writer Charles Frankel's *The End of the Dinosaurs*. Walter Alvarez was in large part responsible for the meteoroid hypothesis and his book was very entertaining. I confess that one of the reasons Frankel's book seemed so special to me is that when I bought it on Amazon it was already out of print, so the copy I received was from the Rockport Public Library with a big stamp labeling it DISCARD. Had the book not been mailed to my house—a far more suitable habitat—it apparently would have gone extinct too.

Both books tell the truly remarkable tale of how geologists, chemists, and physicists established that an enormous meteoroid (remember I'm using "meteoroid" for large objects too) was the most likely cause of the extinction that wiped out the dinosaurs—along with a good deal of the other species that were alive at the time. Evidence abounds that this meteoroid precipitated the dramatic shift in fossil record during the K-Pg transition. All the features that characterize impact craters, including spherules, tektites, and shocked quartz, were found in the vicinity of a boundary iridium layer, which separates relics of abundant life underneath it from the much sparser fossil record above.

The books also relate the incredible and inspiring detective story of how scientists actually found the crater that corresponded to that meteoroid hit, though consultation with experts taught me that some of the literature is a little misleading. I will do my best to get it right here. It's a great story.

Although the idea of meteoroids causing extinctions took hold only in the late twentieth century, people had speculated about their potentially dire consequences for centuries. When people first noticed comets, they considered them life threatening—but for superstitious and unsubstantiated reasons. In 1694, Edmond Halley boldly suggested a comet was the source of the biblical deluge. About 50 years later, in 1742, the French scientist and philosopher Pierre-Louis de Maupertuis put the potential threat from comets on stronger scientific footing when he recognized that the disturbances to the ocean and atmosphere that a comet strike could create had the potential to wipe out many forms of life. Another Frenchman, the great scientist Pierre-Simon Laplace, whose work on the Solar System's formation survives to today—also suggested that meteoroids could precipitate extinctions.

But their ideas were largely ignored, since they couldn't be tested and furthermore seemed a little crazy. Neglected too were the ideas of the American paleontologist M. W. de Laubenfels, who in 1956 recognized the potential import of the meteoroid that hit Siberia in

1908 and devastated a vast swath of forest—identifying the damage such as fires and heat that even a fragment of a comet could cause. In an amazingly prescient analysis, he also recognized that these environmental impacts would affect various species differently, so that burrowing mammals might survive—as indeed turned out to be the case following the K-Pg event.

Even as late as 1973, most scientists ignored the geochemist Harold Urey when he suggested, based on the glassy tektites of molten rock, that a meteoroid impact was responsible for the K-Pg extinction. Urey was a little overly enthusiastic, however, in suggesting that not only the K-Pg extinction, but all other mass extinctions, were due to comet impacts. Even so, he anticipated future studies and helped turn earlier proposals into real science by pointing out that detailed investigations could identify rocks whose shape or composition could be explained only by the heat and/or pressure of a meteoroid hit.

However, any such smart and prescient ideas were essentially ignored before Alvarez made his proposal. The notion of a cosmic impact causing an extinction was radical even in the 1980s, and might have sounded somewhat absurd on first hearing. It is reminiscent of some theories I hear from 12-year-olds who attend my public lectures, where they try to impress me by combining all the scientific terms they have ever heard. This can lead to contrived and usually pretty funny scenarios, such as when one youngster asked me about a theory he claimed he had always wondered about in which black holes from warped extra dimensions solved all the remaining problems of the Universe. Fortunately, he laughed when I suggested to him that he hadn't actually always thought about this.

But as with any more radical theories that eventually take hold, the meteoroid proposal could explain observations that defied more conventional explanations. No terrestrial process could account for all the detailed phenomena that would be found that ultimately supported the hypothesis. The proposal gained credence because it made predictions, many of which have since been validated.

HOW INSPIRATION STRUCK

The tale of Walter Alvarez's scientific sleuthing began in Italy. The hills of Umbria near Gubbio, a couple of hundred kilometers north of Rome, reveal a marine sediment that dates from the Late Cretaceous to the Early Tertiary (now the Paleogene). The Scaglia Rossa, as it is known due to its pink color, is a sedimentary rock consisting of a very unusual deep-water limestone—calcite or calcium carbonate, which is what most seashells are made of and is what bone supplements sometimes contain—that was formed on the seafloor and later pushed upward so it is now exposed. This meant that the evidence of the extinction—a thin layer of clay separating the lower, whiter rock layer from a red layer above—would have been visible to an attentive passerby. The fossils in the lower, whiter rock are mostly the remnants of foraminifera, single-celled protozoa that live in the

[FIGURE 30] With the Geoparque director, Asier Hilario, at the K-Pg boundary at Itzurun Beach in Zumaia, Spain. (Jon Urrestilla)

deep oceans and are extremely useful to us for deducing the ages of sedimentary rocks. But only the very smallest of the foraminifera are present in the upper, darker layer. The "forams" almost went extinct along with the dinosaurs, making the extinction boundary very clear.

The Flysch Geoparque that I visited during my recent university visit to Bilbao contains a piece of the K-Pg boundary—which appears as a thin dark line near the limestone cliff's base. Like everywhere else on the globe where this clay layer exists, the boundary dates from the time of extinction. I considered myself very lucky when my physicist colleague and his geologist cousin helped arrange a visit to the incredibly beautiful site at Itzurun beach, where I could run in at low tide to view the boundary up close. Getting to touch this 66-million-year-old piece of history was almost surreal. (See Figure 30.) Though the cliff derives from the distant past, its trove of information is still with us as part of our world.

AT THE K-PG BOUNDARY

In the 1970s, Walter Alvarez studied a similar boundary layer in the Scaglia Rossa, where he focused his attention on the clay that separated the lighter-colored limestone below that was full of fossils from the darker limestone without them above. This clay that Alvarez made the target of his studies was critical to unraveling the cause of the devastation that took place 66 million years in the past. The thickness of the clay depended on the length of time that passed between the deposition of the lighter and darker rock and could therefore help him determine if the extinction event was rapid or slow.

When Alvarez first started thinking about the K-Pg layer in the 1970s, geology was dominated by the uniformitarian, gradualist viewpoint that had recently been vindicated by the theory of plate tectonics developed over the previous two decades. Entire continents could gradually move apart, mountain ranges could form over time, and canyons as deep as the Grand Canyon could emerge through gradual effects—including a river such as the Colorado River cutting through the ground, water- and ice-induced erosion, land plate movements, or magma eruptions—any of which could dramatically alter terrain over time. No catastrophe was needed to account for these seemingly very dramatic changes.

The limestone formation seemed mysterious in that the difference between the upper and lower levels of limestone indicated a very abrupt transition, inconsistent with the gradualist point of view. Had he been there, Charles Lyell would have interpreted the thinness of the K-Pg layer as simply misleading and concluded that despite appearances, it had taken many years to create. Darwin might have thought the formation was simply an illusion created by an inadequate fossil record.

The only way to truly know if the transition was sudden—and not simply a clay deposit that had washed in over a few days—was to measure how long it would have taken to deposit the clay that separated the two differently colored limestone layers. And that was the task Alvarez, who had a long-standing interest in dating geological events, set for himself. Alvarez hoped to study geomagnetic reversals to learn more about the timing of the deposition of the K-Pg boundary, which he knew could be an important clue to the triggering event. (Andy Knoll, a professor of natural history and Earth and planetary sciences at Harvard, mentioned that Alvarez and his wife might have been even more interested in looking at Medieval art and architecture. I suspect both interests played a role.)

But the better method for measuring how long it took for the clay to be deposited turned out to be measuring its iridium content. Iridium is a rare metal and, next to osmium, is the densest element. Its corrosion-resistant properties make it useful for spark plug electrodes and fountain pen nibs, among other things. It also turns out to be good for science. The iridium spike that Walter Alvarez and his collaborators discovered turned out to be the key to nailing down the origin of the extinction event.

Although I had known for a while about the iridium spike, I was astonished when I more recently learned that the original intention of Walter and his physicist father, Luis Alvarez, had been to measure iridium levels in the clay based on reasoning exactly opposite to what they soon realized was true. Luis Alvarez knew that meteoroids had a much higher level of iridium than the Earth's surface. Although the iridium level on Earth should be the same as that of a meteoroid, most of the Earth's original iridium long ago dissolved into molten iron and sank with it to the Earth's core. So any iridium on the surface should have an extraterrestrial origin.

Luis Alvarez assumed that meteorite dust should be deposited at a fairly steady rate. (Actually he had originally suggested using beryllium-10, but the half-life turned out to be too short to be a practical tool for this problem.) Iridium levels on the surface should be quite low were it not for the deposition from this steady extraterrestrial "rain." The two Alvarezes had the clever idea that by studying iridium levels on Earth, they could access this cosmic hourglass that would help determine how long it took to deposit the K-Pg boundary clay. What they expected was a smooth distribution over time indicative of a steady, nearly constant deposition that could be used to deduce the length of time it took for the clay layer to form.

What Walter and his collaborators found when they examined the actual rock was completely different. The surprise that convinced Alvarez that something odd was afoot (in his case quite literally) came from much higher than expected iridium levels in the clay. In 1980, a team of scientists at the University of California, Berkeley—the father-son team of Luis and Walter Alvarez, partnering with the nuclear chemists Frank Asaro and Helen Michel, who could measure iridium abundances at extremely low levels—found a decisive elevation in iridium—30 times higher in the Scaglia Rossa than in the surrounding limestone. And that number was later corrected to 90.

This type of formation was found not only in Italy (sadly so many people have sampled the Scaglia Rossa since Walter Alvarez's time that the K-Pg boundary clay is now hard to reach), but all over the globe, and iridium levels in those locations noticeably spiked too. In a similar clay layer in Stevns Klint—a coastal cliff with well-preserved K-Pg evidence in Denmark—the increase was by a factor of 160. Other laboratories confirmed the elevated levels of iridium in similar boundary layers elsewhere.

If the original hypothesis (and incentive for the measurement) was correct and meteoritic dust rained down at a constant rate, the K-Pg clay would have required more than three million years to form. But that was far too long a time for the thin clay layer that represented the K-Pg boundary. Alternatively, if the iridium level was similarly elevated all around the globe, then 500,000 tons of iridium—thought to be a rare metal on Earth—had suddenly descended on our planet at the time of the K-Pg extinction. The only explanation for this enormous deposition could be a cosmic origin. The Earth is so intrinsically low in iridium on the surface that without some extraterrestrial phenomenon the high iridium level would be essentially impossible to explain.

The Berkeley team also determined the relative abundances of other rare elements so they could further narrow down the possibilities. For example, maybe the extraterrestrial source was a supernova. In that case, the supernova would have led to plutonium-244 in the clay too. And the initial analysis made it appear that indeed, this element was present as well. But in a responsible exercise in scientific standards, Asaro and Michel repeated their analysis the following day and discovered there was no plutonium to be found. The initial finding was simply contamination in the sample.

After racking their brains for alternatives, the Berkeley scientists were left with basically one plausible explanation for the high levels of iridium—a large impact from an extraterrestrial object that occurred roughly 65 million years before. In 1980, the group led by Walter and Luis Alvarez proposed that a big meteoroid had collided with the Earth and rained down rare metals, including iridium. Such a meteoroid impact—either by an asteroid or from a comet—was the only event that could account for both the total amount of iridium and the elemental ratios, which matched those that are characteristic of the Solar System.

Based on the measured iridium and the average meteorite iridium content, the researchers could furthermore guess the size of the object that had hit. They concluded that it had to have been an incredible 10–15 kilometers in diameter.

STRIKING EVIDENCE

Given the many fatal mechanisms that a huge meteoroid provides, as well as the paucity of adequate explanations for the geological evidence associated with the K-Pg extinction, an extraterrestrial explanation seemed a plausible and reasonable alternative to more conventional suggestions such as geologically or climate-induced processes. Yet despite the compelling nature of the hypothesis, any scientist—no matter how daring—has to exercise caution when introducing a new idea. Sometimes radical theories are correct, but more often than not, a more conventional explanation has been overlooked or not properly evaluated. Only when existing scientific ideas fail where more daring ones succeed do new ideas get firmly established.

For this reason controversy can be a good thing for science when considering a (literally) outlandish theory. Although those who simply avoid examining the evidence won't facilitate scientific progress, strong adherents to the reigning viewpoint who raise reasonable objections elevate the standards for introducing a new idea into the scientific pantheon. Forcing those with new hypotheses—especially radical ones—to confront their opponents prevents crazy or simply wrong ideas from taking hold. Resistance encourages the proposers to up their game to show why the objections aren't valid and to find as much support as possible for their ideas. Walter Alvarez even wrote that he was pleased it took a while for the meteoroid idea to find conclusive support since it allowed time for all the secondary evidence that strengthened the case to be found.

The meteoroid hypothesis was indeed met by resistance from those who thought it an extravagant theory—many of whom preferred the gradualist point of view. Confusingly, plate tectonics lent support to this viewpoint at about the same time that the Moon missions, with their close-up views of many craters, were making a strong case for the possible catastrophic effects of impacts. Perhaps these two different advances were why—taken as a group—geologists tended toward the gradualist viewpoint whereas physicists went for the catastrophic.

Of course the Moon's craters could have all been created in the early stages of its formation—and in fact most of them were—so their existence was not in itself an argument for the significance of meteoroid impacts in later evolution. Still, their prevalence should have made less surprising an assumption that not only gradual, but also catastrophic, processes played a role in our Solar System and in the development of its life. Craters were clear and palpable evidence of lunar impacts. The Earth is both bigger and very near the Moon so clearly meteoroids would have hit here too.

But at the time of the Alvarez proposal, many paleontologists favored gradualist explanations. Some took the perspective that dinosaurs had simply died away in the late Cretaceous period due to some form of adverse environmental conditions—such as climate change or bad diet. Many others thought volcanic activity was the culprit. Support for this point of view came from the Deccan Traps in India, which were formed by an enormous amount of volcanic activity that occurred around the time that the dinosaurs went extinct. The Deccan Traps cover a region bigger than half a million square kilometers—comparable to the size of France—and they are about two kilometers thick. That's a lot of lava. To further confuse the situation, the traps can be dated to very near to the Late Cretaceous–Early Tertiary boundary.

Indeed, groups of dinosaurs such as sauropods, a group that includes the apatosaurus—the original and perhaps temporarily preferred name for the brontosaurus (a debate rivaling that over planet Pluto)—had already gone extinct by the end of the era. But support for the gradual decline idea stemmed in part from the incomplete fossil record that existed at the beginning of the investigation, which became less compelling as additional regions were studied and more fossils were found. Fossils discovered in Montana revealed at least between 10 and 15 dinosaur species that survived up to the end of the Cretaceous period. Recent digs in France uncovered evidence of dinosaurs within a meter of the K-Pg boundary and those in India showed dinosaur evidence below the boundary as well. Other species, such as ammonites, did show a decline in diversity at first. But closer and broader inspection again revealed that at least a third of the dinosaur species survived to the boundary—though some had indeed gone extinct earlier.

On top of that, although people initially thought that the traps were created very quickly, later work showed that their formation took a few million years, and that the K-Pg event corresponds to a layer in the middle, which oddly enough seemed to be a time of suppressed volcanic activity. Probably the most convincing evidence that the volcanoes were not solely responsible for the dinosaurs' extinction is that Indian geologists have found dinosaur bones and fragments of their eggs right in the sediment up to the region that constitutes the K-Pg boundary. Dinosaurs were not only alive—they were living in the region of the traps themselves.

Even so, more recent developments place the formation of the traps closer to the extinction time than people had formerly thought, supporting a place for some volcanic activity—even if not for all the destruction. Some speculate that the volcanic activity was in fact a result of the meteoroid impact, in which case whatever volcanic effects took place were indirectly attributable to the meteoroid too. Whatever their role, volcanoes don't explain the many other coincidences in geological features that persuasively argue for the significance of the meteoroid.

Indeed, once people started to look in earnest, evidence for the meteoroid hypothesis rapidly accumulated. Details matter and they can help resolve many a controversy. After the 1980 Berkeley suggestion, the K-Pg clay layer was meticulously studied in Italy, Denmark, Spain, Tunisia, New Zealand, and the Americas. By 1982, almost 40 locations around the globe had been carefully scrutinized. The Dutch paleontologist Jan Smit observed high iridium levels in Spain, and other paleontologists measured them in Stevns Klint. Smit also measured other rare metal levels, such as those of gold and palladium. He found osmium and palladium levels a thousand times higher than seen elsewhere on Earth. And again, the relative metal abundances corresponded to those expected in meteoroids.

Some scientists in favor of the volcano explanation suggested that large amounts of iridium were pumped up by volcanoes from the

Earth's mantle and core, where levels are known to be higher. But known volcanoes don't emit nearly enough iridium to account for the 500,000 tons worldwide that Alvarez and others calculated to be present at the K-Pg boundary, even allowing for other potential concentrating effects such as precipitation in the ocean. In any case, iridium isn't the only heavy element in meteorites and the abundances of other elements didn't match those in volcanic emissions either.

Supplementary support from other observations in and around the K-Pg layer provided further evidence of the meteoroid proposal. The discovery in diverse locations of rock droplets such as *microkrystites*—smaller versions of tektites, the glassy rocks with rounded shapes that emerge from impact melts that have spun and solidified in the atmosphere before dropping back to Earth—lent support to the meteoroid idea.

But these glassy spherules, as they used to be known, also presented a red herring at first. The chemical composition resembled that from ocean crust, which it turns out was very likely representative of the impactor and not the target. Had the initial misleading conclusion been correct and the landing been in the ocean and not on land, it would have meant that despite all the mounting evidence for an impact, the impact site would probably have remained concealed.

This misplaced concern abated when geologists found evidence indicating that the meteoroid had landed on a (potentially accessible) continental shelf. This was the discovery of shocked quartz, which indicates a high-pressure origin that could come only from collisions on quartz-containing rock. Rocks that don't melt are shattered, so the minerals they contain can move to form crisscrossing bonds. (See Figure 24.) The only known sources for these bonds are meteoroid impacts and nuclear explosions. Presumably nuclear tests didn't happen 66 million years ago—although a researcher told me a radio interviewer actually asked him about this possibility—which left a meteoroid impact as the only potential surviving explanation.

In 1984, when shocked quartz was found in Montana, and later in New Mexico and Russia, the new discoveries argued strongly for a meteoroid impact too. That this evidence was a type of quartz argued furthermore that the crater, assuming there was one, should be located on land, since quartz is rare in rocks from the ocean.

Further evidence continued to accumulate in the meteoroid hypothesis' favor. Canadian scientists found microscopic diamonds in the K-Pg layer in Alberta. These could have come from meteoroids that simply couriered the diamonds from outer space, or they could have been formed on impact. Detailed studies of size and carbon isotope ratio favored the latter interpretation. In Canada, as well as in Denmark, amino acids that are not known to exist anywhere else on Earth were discovered in the layers. This evidence had the interesting feature that it favored a comet interpretation, since these amino acids were found in the bordering limestone as well—as would be the case if comet dust were around at the time the layer was formed.

One other important geological feature that argued for high-pressure impacts were crystals called *spinels*. These are metal oxides that contain iron, magnesium, aluminum, titanium, nickel, and chromium and which have bizarre snowflake, octahedral, or other shapes that indicate rapid solidification after high-temperature melting. Spinels occur in volcanic magma too, but the spinels that were found contained the elements nickel and magnesium—unlike volcanic spinels, which contain more iron, titanium, and chromium. Better yet, the amount of oxygen helps determine where the spinels were formed. The oxidated spinels of the K-Pg layer indicated a low-altitude origin—below 20 kilometers. The crystals also were found only in a thin layer, confirming the hypothesis that the catastrophic event that occurred at the K-Pg boundary was very brief.

Volcanoes couldn't account for the shock-induced materials. Although volcanoes do produce deformations, existing volcanic regions don't produce the shocked quartz that would be necessary to match observations from the time of extinction. The dislocations in volcanic shocked quartz run along a single plane, rather than two or more intersecting ones, which is a phenomenon that occurs only at high shock pressures. These details are important since these phenomena are found in precisely the location that demarcates the K-Pg extinction boundary.

Even so, having firmly established the meteoroid's destructive influence, we shouldn't entirely dismiss the gradualist viewpoint. Very likely, conditions were changing around the time of the K-Pg extinction in a way that increased the fragility of the ecosystem so that when a meteoroid did strike, it did more damage than it would otherwise have. Evidence shows that some significant fraction of species had gone extinct even before the more dramatic extinction event took place. Recent more precise measurements of the timing of the Deccan Traps lend further support to volcanic activity playing some role. Though unlikely to be responsible for the extinction event that eventually happened, volcanoes and other phenomena very likely abetted it—both before and after the meteoroid struck.

But to do colossal damage, the impact didn't need any assistance.

HOW LIFE STRUCK OUT

It's difficult to fathom just how huge and devastating the meteoroid must have been. The impactor had a span that was about three times the width of Manhattan. And it wasn't just big. It was also moving very quickly—at least 20 kilometers *per second* and if a comet, perhaps three times faster. The object's speed was at least 700 times faster than a car on a freeway moving at 100 kilometers *per hour*. This impactor would have been an object the size of a major city moving 500 times faster than a vehicle on an autobahn. Since the energy an object carries increases with its mass and with the square of its speed, such a fast, big object hitting the Earth would have had a devastatingly enormous impact.

To put it in some perspective, an object of its size and speed would have released an energy equivalent of up to 100 trillion tons of TNT, more than a billion times greater than that of the atom bombs that destroyed Hiroshima and Nagasaki. The comparison is not accidental. Luis Alvarez worked on the Manhattan Project and he made similar observations. More broadly speaking, the Cold War preoccupation with the effects of nuclear explosions enhanced people's interest in the crater. Research on both benefited from increasing knowledge about the K-Pg impactor's long-term environmental effects.

The Tunguska object and the meteorite that created the Meteor Crater in Arizona carried a fraction of this energy—the equivalent of perhaps 10 megatons of TNT. The diameter of the impactor in each case was more like 50 meters, rather than 10 to 15 kilometers for the impactor creating the K-Pg event. Krakatau's energy was only a few times more than that of these smaller meteoroids, which was comparable to the most powerful nuclear weapons ever made (about 50 times what exists now). A kilometer-wide meteoroid would already be sufficiently big to do global damage. The object Alvarez suggested was at least 10 times bigger than that—bigger than the height of Mount Everest, which stands nine kilometers above sea level.

The impact (in both senses of the word) of this enormous speeding object was devastating. As described in Chapter 11, many disasters follow such a huge heavy rock hurling itself into Earth. Near to the blast—within about a thousand kilometers—extreme winds and waves raged, and huge tsunamis radiated from the blast site. These tidal waves would have been enormously powerful, though of limited range since as it turned out the water depth back at the impact location was only about a hundred meters. Tidal waves would also have appeared on the opposite side of the globe, triggered by perhaps the most massive earthquake the Earth has ever experienced. Extreme winds would have blown outward from the impact location, and then rushed back in. This would have carried a cloud of superheated dust, ash, and steam that was thrown up when the meteoroid initially drove into the ground. This wind and water would have used up about one percent of the impact energy. The rest would have gone into melting, vaporizing, and sending seismic waves throughout the Earth—the equivalent of 10 on the Richter scale.

Trillions of tons of material would have been ejected from the site of the crater and distributed everywhere. Afterward, when the hot solid particles descended through the atmosphere, they would have been heated to incandescence and raised temperatures around the globe As a consequence, fires would have raged everywhere and the Earth's surface would have literally been cooked. In fact, in 1985, the chemist Wendy Wolbach and her collaborators found evidence of fires in the K-Pg layer in the form of charcoal and soot. The abundance and shape of the carbon flecks they found confirmed that the fires happened—and destroyed then-existing plant and animal life. The researchers concluded that more than half the world's biomass was incinerated within months of the impact.

And that's not all. The water, air, and soil were all poisoned. Perhaps people weren't just superstitious in their fear of comets, which turn out to contain poisonous materials such as cyanide and heavy metals including nickel and lead. Although some chemicals would have vaporized before they could do any damage, very likely heavy metals rained down from the sky.

Probably even more damaging would have been the nitrous oxide created in the atmosphere, which would have descended to the ground in acid rain. Sulfur would have been released into the atmosphere too, creating sulfuric acid that could have remained there and blocked sunlight, creating global cooling that followed the global heating immediately after the catastrophe occurred and lasting perhaps for years. The loss of photosynthesis would have reverberated throughout the food chain. Global warming and dust particles blanketing the Earth could have played some role too—extending the deviant heating and cooling for many more years.

Indeed, fossil records demonstrate that the destruction's legacy persisted well after the initial impact. Even the species that did survive had their ranks severely diminished. The oceans didn't recover for hundreds of thousands of years and most likely saw destructive influences for at least a half million to a million years afterward. The fossil record demonstrates the absence of plankton and other fossils in the darkness of the limestone that contains little or no carbonates. There is instead evidence primarily of detrital particles—the small fragments of weathered and eroded rock that remained. The normal color doesn't resume for at least centimeters and sometimes meters in these layers, depending where on the globe one looks.

The many disasters provided abundant opportunities for plants and animals to go extinct. It seems no living creature survived that was heavier than about 25 kilograms—about the weight of a mid-size dog. In order to make it through, a means of hiding from the disaster—through hibernation or otherwise—would have been essential. Depending on their reproduction methods (seeds had a better chance of surviving than other means of procreation) and on their source of food (species that fed on waste fared better) some species did survive. Animals that could escape into the skies had better chances too. But most plants and animals died. A 10- or 15-kilometer-wide meteor was enormously devastating—to the environment and to life.

STRIKING PAYDIRT: REDISCOVERING THE CRATER

Nonetheless, researchers at the time knew that even with all the evidence discovered in the 1980s and the increased understanding of the implications that a huge meteoroid could have on life on the planet, finding a tangible 66-million-year-old crater of the right size would clearly strengthen the case for the impact suggestion. A crater would not only substantiate the hypothesis, but would permit more detailed investigations that could better pinpoint the size and time of the meteoroid hit, along with other features that could help confirm an impact.

The crater's size—along with its age—was a critical prediction. Based on the amount of iridium that had been measured, Walter Alvarez deduced that the meteoroid should have been at least 10 kilometers across, so the crater should be about 200 kilometers wide, since craters are usually about 20 times the size of the impacting object. Alvarez wasn't the only one to estimate a crater of this magnitude. Another paleontologist independently predicted a size of 180 kilometers based on the assumption that the clay contained seven percent meteorite material, with the rest just pulverized rock from the target.

A crater of the right scale from the correct date would be the smoking (technically no longer smoking) gun for the Alvarez proposal. Yet it took more than a decade for the crater to be discovered—spawning one of the best detective stories in modern science. In fact, the odds of finding the impact site didn't seem all that favorable when people first began to look. Although some large craters have been discovered over the years, many more go missing. Even if we are "lucky" enough that a meteoroid hits land and not the ocean, erosion, burial by sedimentation, or tectonic destruction can eliminate any sign that a crater had been formed.

For the meteoroid responsible for the K-Pg event, the discovery challenge was exacerbated by the apparent lack of clues to the strike's location. The very ubiquity of the iridium and other geological evidence, distributed more or less uniformly around the globe, confirmed the meteoroid's worldwide impact but didn't single out any particular region. When people first began to search, it seemed a daunting, if not impossible, task to determine where on Earth one particular meteoroid had hit more than 65 million years ago.

However, in the crater hunter's favor, the shocked quartz that had been found suggested its origin was continental—or on a continental shelf—so that searches on land stood a chance of successfully identifying the remains of the culprit. Several seemingly promising candidate craters emerged but were soon ruled out on further investigation—disagreeing with more precise measurements of their time of impact, determinations of their size, or mineralogical studies.

But one very important independent observation had been largely overlooked for quite some time. As early as the 1950s, industrial geologists had identified a buried circular structure 180 kilometers in diameter that extended half onshore, under the Yucatán limestone plains, and half offshore, where it was buried under water and sediment in the Gulf of Mexico. Geologists from Petróleos Mexicanos, or Pemex, as the Mexican oil company is known, drilled wells into this feature. They hit crystalline rock at a depth of about 1,500 meters, leading them to think they had found evidence of a volcano rather than what for them would have been a far more interesting oil trap.

But in the late 1960s, the geologist Robert Baltosser—who was involved in a second round of exploratory drilling, just in case the first investigators had missed an oil cache—suggested that it could in fact be an impact crater. His suggestion was based on measurements of the shape of the feature's gravitational potential—how the force of gravity varied over the circular structure. But it still wasn't oil and Pemex didn't allow him to release his observations. The consequence was that most of the people who knew about the structure worked for the oil industry, which for obvious reasons did detailed surveys of the ocean floor but wanted to protect their results.

But Pemex was persistent in its oil search and in the 1970s did further geological studies, including an aerial magnetic survey over the entire Yucatán Peninsula. The American consultant Glen Penfield noticed a strong magnetic anomaly about 50 kilometers across bordered by an outer ring with uncommonly low magnetism that was about 180 kilometers in diameter. This is precisely the pattern expected for a large impact crater, with the central region associated with the impact melt and the outer region containing hardened target debris. This correspondence wasn't lost on Penfield. Aerial gravity data lent further support to this interpretation. The heightened and depressed gravity field correlated with the variations in the magnetic signals.

So as early as 1978, Penfield had a reasonably strong indication of an impact crater. He recognized that evidence for a previously unknown impact event could be a pretty big deal so he got permission from Pemex to release what was normally considered to be proprietary data. Along with Pemex geologist Antonio Camargo, Penfield presented his results in 1981 at the Society of Exploration Geophysicists convention in Los Angeles. But the discovery didn't get a lot of attention. Most people who were listening were still unaware of the impact hypothesis for the K-Pg extinction so no one at the time envisioned such a connection.

In fact, most people interested in locating the impact crater responsible for the K-Pg extinction didn't get around to studying this particular crater until 1990. But how they arrived at it is also an incredible story. Those in search of a specific 66-million-year-old crater of about 200 kilometers in diameter to verify the Alvarez proposal had gone about the search from an entirely different perspective than the Pemex geologists. They studied the K-Pg layer searching for hints of the location of an impact. Despite the uniformity of the iridium deposits across the globe, they knew of one clue that—if found—promised to be more location specific. If the meteoroid had hit the ocean but landed close to shore, it would have created a tsunami powerful enough to leave its trace in the continental platform. This might have seemed like wishful thinking, given the evidence of a terrestrial hit, but geologists kept their eyes peeled and were rewarded for their efforts.

In 1985, Jan Smit and a collaborator studied an outcropping of disturbed sediments in the K-Pg sediment in the Brazos River bed in Texas near the Gulf of Mexico, which they were convinced had been shaped by the proposed tsunami. The geologist Joanne Bourgeois of the University of Washington carefully followed up their work, finding unusually coarse sandstone containing fragments of shells, fossilized wood, fish teeth, and clay matching the local seafloor—and fixing the location's depth 66 million years earlier to be 100 meters below the level of the sea at that time. She was able to use the size of the sandstone blocks to estimate that the current ran faster than a meter a second, corresponding to a wave height of at least 100 meters, and furthermore found clay with patterns indicative of a current that moved both to and from shore. By assuming the maximum possible size wave as being the same as the full sea depth of 5,000 meters, Bourgeois deduced that the impact must have been less than 5,000 kilometers away from her site, which meant in the Gulf of Mexico, the Caribbean, or the western Atlantic.

The other hint of the location came from the geologists Bruce Bohor and Glen Izett, who in 1987 found that the largest and most abundant deposit of shocked quartz was found in the western interior of North America, suggesting impact near the continent. This was consistent with the analyses of Smit and Bourgeois, which had suggested that the impact was near the southern end of the continent.

The impact site was narrowed down even further when the Haitian geologist Florentin Maurrasse identified some interesting debris at the K-Pg boundary layer in his native country. His description of the unusual sediments attracted the attention of the University of Arizona graduate student Alan Hildebrand, his thesis advisor Bill Boynton, and the researcher David Kring. Although Maurrasse had described the debris as volcanic in origin, the Arizona group were aware of how readily confused volcanic and impact debris can be. Once they saw the Haitian samples, they recognized the tektites and decided to visit Haiti themselves. There, in 1990, they found a half-meter-thick sediment outcrop that seemed to contain tektites—and shocked quartz and iridium clay as well. This had the appearance of a region that was very likely associated with the meteoroid impact. From the thickness of the layer, they concluded that the crater should have been no farther than about a thousand kilometers away at the time of impact.

Though Hildebrand initially favored a possible Caribbean candidate that he later rejected, the Arizona team ultimately zeroed in on the Yucatán feature that had been identified a decade before. Yet it was not scientists, but a reporter—Carlos Byars of the *Houston Chronicle*—who first made the connection. After listening to Hildebrand present the Arizona group's research at a scientific meeting, Byars told him about Penfield's earlier discovery of a potential impact crater—helping the scientists bring the mystery of the missing crater to its remarkably satisfying conclusion.

The Pemex-discovered crater was in the right location. And it also had the right size. This agreement was a major argument in favor of the crater's connection to the K-Pg extinction. Even so, when in 1990 Hildebrand submitted two abstracts suggesting the connection to a scientific journal, they went unpublished—in part because the initial evidence wasn't sufficiently convincing. However, opinions changed when the Arizona team identified shocked quartz from the crater.

Because the crater was in a submerged continental platform, sediment covered it, making the crater hard to find and study. But its burial was fortunate in some respects too, since the thousand meters of hardened mud above it shielded the crater from the erosion that would have occurred had it been on the surface. In order to investigate the buried and therefore initially inaccessible crater, the Arizona scientists contacted Penfield and Camargo to study the cores that had been drilled before. They obtained two thumb-sized samples stored in New Orleans. Sure enough, when the Arizona group studied these old Pemex drill cores, they found what they were looking for. They identified shocked quartz and impact melted rock that demonstrated the crater came from an impact and not a volcano. Kring announced the discovery at the NASA Johnson Space Center in March 1991.

The Arizona scientists combined their drill core studies with the geophysical data that Penfield and Camargo provided and the groups—along with a couple of other collaborators—accumulated strong evidence that the crater was the result of the impact that precipitated the K-Pg mass extinction. They presented this result, together with their size estimate of 180 kilometers, in the journal *Geology* in 1991. Confronted with the shocked quartz and other supporting evidence, many scientists began to pay attention.

The Arizona team named the crater after an unfortunately hard-to-pronounce small nearby fishing harbor, Chicxulub Puerto, which is located above the center of the structure. The term, which is pronounced CHICK-shuh-lube, is sometimes translated as the devil's tail—appropriately enough for the imposing feature that Walter Alvarez dubbed the "crater of doom."

Soon after the Arizona team's publication, remote-sensing experts realized they could detect the crater's circumference in satellite imagery, where small ponds were in a ring around the crater, 80 kilometers in radius. These were also most likely caused by the crater's formation, which would cause groundwater to swell upward and cut through the Earth's surface, and was therefore further evidence of the crater connection.

More support followed suit. The Arizona scientists could tell the material in the older cores was impact melt and that it contained features that resembled the microtektites of the K-Pg sediment in the surrounding gulf. Kring and Boynton also observed chemical similarities between the Chicxulub melt rock and glassy spherules deposited at the K-T boundary in Haiti—solid evidence that the Chicxulub crater was produced precisely at the K-T boundary where life gets extinguished. The evidence had become so strong by this point that the discovery made headlines and entered the public domain.

Geologists went on to find still more links between the Yucatán crater and the K-Pg extinction. Near to the crater in precisely the right boundary region, Jan Smit and Walter Alvarez identified precisely the sort of geographical outcrop—a tumble of breccias containing spherules and even glass. The glass discovery was important too. Glass forms only during a quick process like an impact and not during a relatively slow one like a volcano, where atoms and molecules have time to crystallize. The glass's streakiness was further indication that it formed too quickly to homogenize.

Exploration and conversations with resident Mexican geologists exposed more regions that had been disrupted by the relatively nearby impact. Studies also showed that the thickness of the boundary ejecta in North America decreased with distance from the site exactly as predicted using Chicxulub as the source. And the geologist Susan Kieffer helped explain the relative distribution of iridium, impact melt, and shocked quartz in terms of successive ejections from the explosion.

By 1992, given all the accumulated evidence, most geologists were convinced that the Yucatán feature was indeed an impact crater. But they still weren't sure about its relationship to the K-Pg extinction. Detailed dating, which would require studying the chemical composition of good-quality cores from the crater, would be the only way to firmly establish this connection.

Scientists managed to successfully determine the age of existing cores—in particular in three well-preserved beads of glass—by studying argon isotopes in the rock. They then dated the spherules from the Haitian K-Pg layer to check whether the impact and extinction times agreed. When the first measurement gave the age as 64.98+/–0.05 million years and the second gave 65.01+/–0.08, the results demonstrated that the events had occurred simultaneously (within the measurement uncertainty). This excellent agreement convinced many scientists that the meteoroid dinosaur extinction theory first put forth by Alvarez and collaborators was correct. The ejecta fell exactly on the paleontological boundary, confirming that the impact occurred at the same time as the extinction.

However, the initial dating of both the crater and the iridium layer—critical to establishing the causal relationship—turned out to be off by about a million years. The relative dates hadn't changed, but the decay constants essential to assigning an age had initially been slightly incorrect. This is why we now think the K-Pg extinction occurred 66—not 65—million years ago.

Even stronger corroborating evidence for the meteoroid hypothesis is the recent significant improvement in the measurement of the alignment of dates. In February 2013, the researcher Paul Renne of the University of California, Berkeley and his colleagues showed that the Chicxulub impact and the mass extinction occurred less than 32,000 years apart, an incredibly accurate measurement for events that took place so long ago. The Berkeley team used argon-argon dating—the technique relying on radioactive argon isotopes that was mentioned in the previous chapter—to show that the impact and the extinction occurred within this very small time interval.

The proximity of dates that they found is almost certainly not mere coincidence and was a remarkable vindication of the impact hypothesis. Though the authors of the paper are careful to point out that the meteoroid event might have been the nail in the coffin of an extinction that had already been precipitated by volcanic activity or climactic change, it is now beyond reasonable doubt that the massive meteoroid strike that created the Chicxulub crater was the crucial trigger.

In March 2010, 41 experts on paleontology, geochemistry, climate models, geophysics, and sedimentology met to review the more than 20 years of evidence for the impact–mass extinction hypothesis that had accumulated by that time. They concluded that it had indeed been a meteoroid impact 66 million years ago that had both created the crater and given rise to the K-Pg extinction, with its most notable victim—the venerable dinosaur. A paper published in the journal *Science* in that year presented a consensus view of the meteoroid as the cause of the extinction. A few months later in the same journal, skeptical paleontologists signed off on a different paper in which they too agreed that a meteoroid had at least been a very significant contributing factor.

The Chicxulub crater is among the largest of the ones found on Earth. The story of its identification was an incredible example of science in action, involving clever inductions, testing and validating bold hypotheses, and explorations in regions as far-flung as Italy, Colorado, Haiti, Texas, and the Yucatán. The meteoroid that struck in the Yucatán had a profound influence on the planet and its life. Its origin and its consequences deftly illustrate the Earth's abiding connections to the Universe.

13

LIFE IN THE HABITABLE ZONE

We've come a long way in our journey towards the proposal that suggests how dark matter and the absence of land-dwelling dinosaurs might connect. We've considered a lot of what is known about the Universe, the matter within it, and the development of structures such as galaxies. Closer to home, we've reviewed the five major extinction events, including the well-studied tale of the K-Pg devastation, and we've investigated the composition of the Solar System with a focus on newer discoveries about asteroids and comets.

But progress in science doesn't involve only the known. Critically, it involves the unknown too. Hypotheses frequently begin as speculative attempts to make sense of marginal but suggestive evidence or—in inspired moments—to synthesize big new ideas. The beauty of the scientific method is that it allows us to think about crazy-seeming concepts, but with an eye to identifying the small, logical consequences with which to test them. We can be lucky and our proposals might point the way forward, but we can also be disappointed when initially promising hypotheses prove incorrect after having first led us astray.

Progress is rarely straightforward. This notion was perhaps overzealously expressed in another context by a friend who skis infrequently but ardently. When I met him on the slopes he described his development as "two steps forward, two steps back." But really, even when he thinks he isn't advancing his technique, his increased time on the snow is bringing him familiarity with a mountain and its terrain that will serve him well in future ski adventures. Indeed, when I encountered him a year later on the same run, his comfort level had noticeably improved.

But the attitude he expressed is one anyone conducting research might recognize at times. Even someone who made no mistakes, worked out all his equations correctly, and properly interpreted the data, might eventually find that the idea he suggested—through no fault of his own—just is not the one borne out in the Universe. Even then, as with skiing, the attempts should at least yield a more intimate understanding of the terrain. Our imaginary researcher can take comfort in the knowledge that he has learnt something from his wrong ideas too—at least the right wrong ideas—even when it didn't always seem so at the time. Making assumptions and finding ways to prove or disprove ideas is, after all, the only way to ascertain their validity. In those wonderful instances when the proposals are fortunate or inspired, research leads to real progress. For a scientist—as for most everyone else too—failures fade when confronted with success.

Pretty soon, this book will address some speculative ideas about dark matter. But this chapter briefly turns to one of the most interesting consequences of the matter whose composition we do know—the development and evolution of life. I'll explain some of the factors that might be important to life's origins, the environmental conditions that can accommodate life, and the possible role of meteoroids in life's development. Although many of the ideas that I discuss are supported by scientific research, some speculative aspects are included as well. These usually concern how crucial any particular feature was to life on Earth or would be to new forms of life that might exist elsewhere.

Our focus below on familiar matter is not to suggest that there aren't many speculative ideas about dark matter too, but I will put these aside for now to return to in the book's final part. Even so, we should not completely neglect life's indebtedness to dark matter and to its role in the creation of our stellar environment—ultimately a result of a dense disk of ordinary matter that precipitated out of a galaxy that was initially seeded by the condensation of dark matter. This scaffolding seeded structure that allowed for the creation of stars and heavy nuclei, which would never have been created in time without dark matter's contribution. Dark matter should also be credited with helping to attract into galaxies and galaxy clusters the heavy elements that supernovae create and that are essential to our planet and to life.

But it was a long way from dark matter haloes to life's creation. The Milky Way disk had to form, and then stars and heavy elements and more complex structures after that. Ordinary matter was essential to all these subtle and complex processes for which our Solar System seems particularly well suited. I can't tell you which of the following speculative ideas about the formation of life are correct. But I can say with certainty that in the coming years, science will make progress.

LIFE'S BEGINNINGS

The origin of life is an extremely challenging problem, especially since no one yet knows what life actually is. I doubt we even would have guessed or figured out the makeup of or the conditions necessary for our kind of life unless we'd been presented with the remarkably complex and unlikely example of life that is already here. But although people are aware of many deep fundamental questions that remain to be addressed, humans repeatedly overestimate the amount that we currently understand. One reason I find anthropic reasoning troublesome is that no one yet knows what might be essential to any possible form of life or even to structures such as galaxies that might support it. I am not as confident as others seem to be that any form of life would be similar to ours.

But before asking questions concerning abstract imaginary lifeforms, we might first want to know how and where the life on our planet began. Did it originate locally or did it come from somewhere else in outer space? Some people speculate that comets or asteroids brought ready-formed life to Earth through spores in a scenario called *panspermia*, others argue that a meteoroid impact helped overcome some barrier to life's formation, and others more conservatively suggest that life on Earth developed without any direct extraterrestrial intervention. This last hypothesis has the advantage that among all the places in the Solar System that we know of, the Earth seems to have the conditions that are most conducive to life's emergence. Though similar environments might exist elsewhere—as far as we know only the Earth has the shallow marine environments, such as lagoons or tidal pools, frozen aqueous solutions, or the surface of clays where chemicals can concentrate and react.

The heavy elements of which life is made certainly came from outer space. Hydrogen was present in the very early Universe, but the other essential elements—carbon, nitrogen, oxygen, phosphorus, and sulfur—arose only because of hot dense stellar synthesis and supernova explosions that happened even before our Sun was born. I was happy to cite this sequence of events during an interview that was requested by the students who were looking for near-Earth asteroids at the Tenerife telescope in the Canary Islands. After presenting their more conventional inquiries, they posed the same quirky question that they told me they ask all their interviewees, "What are the properties I think that students and young stars have in common?" I was relieved when the interviewers were satisfied with my response, which was that students absorb ideas and process them to create new ones that they then send out into the world to restart the cycle—much as stars absorb interstellar material to create heavy elements, which they then eject back into space to be processed anew. When molecular material is expelled, dispersed throughout the interstellar medium, and aggregated in dense clouds where some fraction reenters star-forming regions, the distribution pattern is not entirely different from the creation, dissemination, and progression of ideas.

But heavy elements had to be further processed before life could emerge. On Earth this occurred when chemicals formed increasingly complex stable organic compounds, which eventually created self-reproducing RNA, then DNA, and later on cells, and subsequently—

much later—multicellular organisms. These are constructed in part from amino acids, the building blocks of proteins. As we come to understand more about what is necessary to develop DNA and RNA and cellular structure, we might better ascertain the extreme conditions that were essential to the origin of life.

One of many interesting questions regarding life's emergence is how amino acids formed in the interstellar medium and elsewhere. In the early 1950s, Stanley Miller and Harold Urey at the University of Chicago did a famous experiment in which they heated a flask of water that was enclosed by a container filled with methane, ammonia, and hydrogen. Their goal was to mimic the primordial ocean in the early atmosphere. An electrical discharge acting on the water vapor played the role of lightning in their artificially created "atmosphere." Miller and Urey successfully produced amino acids with their simple apparatus, demonstrating that the production of amino acids in solar and extra-solar environments is actually not so surprising.

The early Earth's atmosphere more likely had carbon dioxide, nitrogen and water, and not the less stable methane and ammonia used in the experiment. But it is interesting that the terrestrial distribution of amino acids is remarkably similar to that produced by the Miller-Urey experiment. The key take-away from their results is probably that forming organic material is relatively straightforward on Earth—as well as anywhere else in the galaxy and the Solar System. Keep in mind that the term *organic* in chemistry simply refers to the presence of carbon, not necessarily to elements of life. The term, though unfortunate, is of course not coincidental in that some (but not all) organic molecules are essential to life as we know it.

Indeed, processes involving carbon occur just about everywhere in the Universe. Stellar outflows, the interstellar medium, dense molecular clouds, and protostellar nebulae all contain organic matter. The region around a star like the Sun creates large amounts of organic matter, as did the cold dense molecular cloud in which it was created. This makes organic synthesis relatively unsurprising but also makes the origin of our life's essential ingredients harder to pin down. Some of it might have originated elsewhere but some scientists believe that a lot of organic matter is likely to prove to have been homegrown—or at least created from material that was first reprocessed in the Earth's mantle before entering the molecules that ultimately contributed to life.

We do know that at least some organic material is delivered to Earth through the impacts of objects within the Solar System. The amount of organic matter inside the asteroid belt seems to be markedly smaller than that outside, which is one reason to suspect that some reasonable fraction of the Earth's organic material was delivered from outer space. Another reason is that although mineral remains from the Earth's early years are sparse, the enormous number of Moon craters and the much larger size of the planet tell us that in its early years many impact events must have happened on Earth too. These very likely delivered substantial amounts of organic material.

Amino acids as well as purines and pyrimidines—also essential for DNA and RNA—are indeed found in space. Asteroids and comets both contain amino acids, some of which are part of life here and some of which aren't found on Earth. At least one discriminator that helps distinguish non-biotic amino acids is *chirality*, or handedness. (See Figure 31.) Only left-handed amino acids are present in life on Earth, whereas amino acids from the outer Solar System contain molecules of both handedness. Handedness has to do with the arrangement of atoms around a carbon atom, which has a distinct directionality—like the different directions your fingers take in your left and right hands. However, at least one study found an excess in asteroid deposits of left-handed molecules in one type of amino acid, muddying the connection of a left-handed excess to life.

A lot of what we know about amino acids in asteroids came from the Murchison meteorite, which fell in 1969 close to Murchison, an Australian town near Melbourne. The Murchison meteorite was a piece of asteroid that originated between Mars and Jupiter. It was a *carbonaceous chondrite* type, which as you might guess from the name contains a good amount of organic molecules, including amino

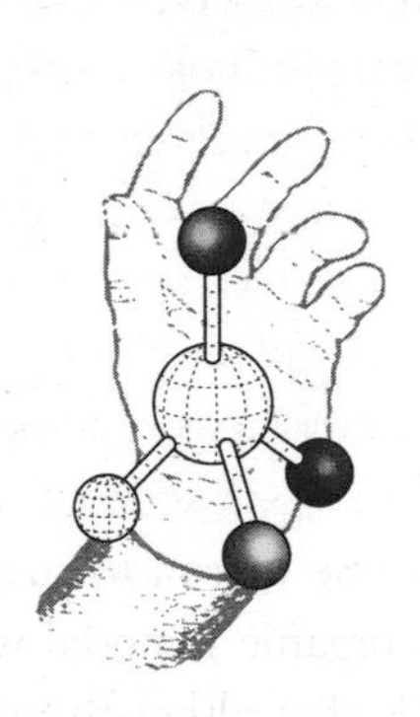

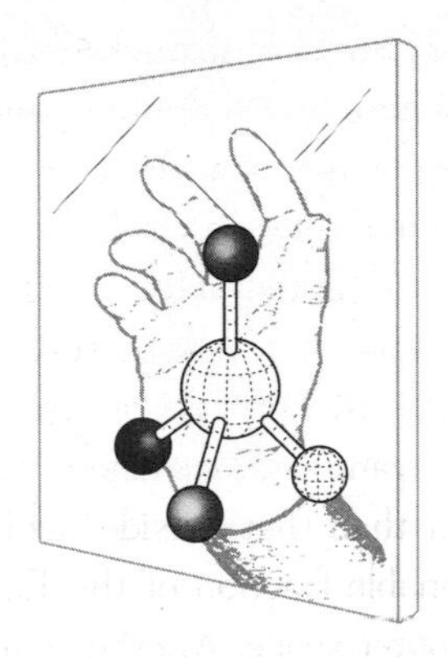

[FIGURE 31] Chiral molecules with a given handedness don't look the same when reflected in a mirror. Amino acids in living beings are left-handed.

acids. By coincidence, laboratories that could study the meteorite had just been built to study lunar samples from the Apollo missions. So scientists fortuitously had the tools to compare the Murchison meteorite to similar ones, such as the Murray meteorite found in Oklahoma, and contrast it with different ones, such as the Orgueil meteorite, which was recovered in France.

Experimenters also try to reproduce cosmic conditions here on Earth in order to study the fate of amino acids that arrived from space. Their research has shown that amino acids can survive comet impacts or be created when extraterrestrial material hits the ground. Observations of the outgassing of comets have demonstrated that whereas most asteroids have highly processed interstellar material, some cometary ices contain early, pristine interstellar matter. The study of meteorites and interplanetary dust, which reflect the content of comets and asteroids that delivered material to Earth, should help establish the origins and amount of some of the molecules that came from space.

Water, like carbon, is probably essential to life in the Solar System—if and where it exists. One particularly notable feature of the Earth is that about 2/3 of its surface is covered by oceans—not all of it and not none of it. This partial coverage by oceans that permits coastlines and tidal regions was probably important to the life that developed here too.

Water is certainly crucial to life as we know it. Evidence in rocks indicates that liquid water has existed in a stable state on the Earth's surface throughout much of its history. Rocks dating back 3.8 billion years ago appear to have formed in water on the Earth's surface. And zircon that dates back well before that—to at least 4.3 billion years ago—exists in a form that seems to have required water in the Earth's early crust.

Life on Earth is certainly greatly indebted to whatever it was that brought to us the vast quantity of water that the planet currently contains. Yet as a friend acknowledged with fascination on a recent ferry ride, the source of this remarkable resource that surrounds us remains a mystery. Some of the water in oceans might have emerged from water trapped in rocks beneath the surface, but the small amount that would have accumulated wasn't necessarily sufficient to account for the large quantities of water that must have been present.

We've already seen that impacts might have delivered organic material that helped facilitate the creation of life. Extraterrestrial delivery of water from comets or asteroids early on—most probably during the Late Heavy Bombardment—is certainly a possibility too. It's a tricky business since most of the water that comes to Earth via meteorites is incorporated into the lattice of its minerals, so some kind of process would have been required to separate it from its mostly silicate host—though some interstitial ice might be delivered by asteroids too.

Comets initially seemed the more likely candidate for water's origins since they are composed largely of ice. However, the isotopes of carbon, hydrogen, and oxygen on Earth don't seem to match what has so far been observed in comets, indicating that comets are probably not the primary source of the Earth's volatiles. This result was substantiated in 2014 when Rosetta returned results indicating that the isotopic composition of hydrogen in the comet it studied didn't match that of the isotopes on Earth—making the hypothesis that most water came from comets even less likely. If objects from space did play a role, most likely the contributions of further-out asteroids, which might have more similar isotope ratios to those on Earth, were important.

A further issue for water in the early Earth is that the young Sun's energy output was probably about 70 percent of what it is today. With the Sun's initially lower luminosity, even the water that did form wouldn't have been in a liquid phase without some other explanation—a quandary known as "The Faint Sun Paradox." However, the young Earth would also have generated heat through the release of gravitational energy as it collapsed, by volcanic activity, via shocks from meteoroids coming through the atmosphere, tidal heating by the Moon which was closer at the time, and by radioactivity from the decay of unstable isotopes inside the Earth. Any of these might have made the Earth warmer than solar radiation alone. Most likely, greenhouse gases, which help warm the planet today, played the most significant role at that time too. Greenhouse gases such as carbon dioxide in the atmosphere cause some of the Sun's light, which hits the Earth primarily in visible wavelengths, to be absorbed in the atmosphere and radiated back in the infrared. Whether or not greenhouse gases fully accounted for the Earth's warmer-than-expected early temperature, liquid oceans were fairly clearly present in the early Earth. So one or more of the above resolutions must have played a role.

THE HABITABLE ZONE

Our cosmic environment contains both friends and enemies—from within and beyond the Solar System. Life seems to rely on a conspiracy of physical conditions in which a suitable ecosystem can flourish—requiring some exceptional conditions that allow it to profit from the beneficial aspects and to deflect or suppress the bad ones. Understanding the prerequisites for life will likely prove to be as daunting as understanding life's origins. But scientists nevertheless hope to determine what makes for a habitable environment—both for elementary microbial life and for the advanced complex life that presumably requires much more particular conditions. Although no one yet has all the answers, anything that makes our environment special merits some attention.

Perhaps it's worth noting that the Sun itself seems special in a few respects. The Sun is among the more massive stars—it's in the top ten percent, might have higher metal content than typical, and is unusually close to the galactic mid-plane for its age. It furthermore seems to have a more circular orbit than similarly aged stars, and is positioned so that it orbits at a somewhat similar rate to the spiral arms and therefore crosses them relatively infrequently. We don't know how essential these atypical properties of the Sun actually are, but any unusual features might be of interest.

Photosynthesis, which relies on solar radiation, is critical to much of the Earth's life. Energy is almost certainly essential to any form of life since it fuels the processes that might create and ultimately sustain it. On Earth, the Sun is undoubtedly the chief source of energy. The power from sunlight today is thousands of times greater than that of the next most significant source, geothermal heat. Even less significant contributions today include lightning—a factor of a million less power—and cosmic rays—over a factor of a thousand lower still.

Although its importance for all forms of life is a matter of speculation, liquid water is certainly important for life that exists on Earth. In addition to knowing where water came from, we'd like to know where its liquid form would be stable. Addressing this question requires not only knowledge about the Sun and our distance from it, but also an understanding of the effectiveness of radiation, other possible heat sources, and the amount of pressure in the atmosphere.

Based solely on the Earth's reflectivity and the Sun's luminosity and distance from us, water on the Earth's surface would be frozen even today without the atmosphere's warming effect. Although we legitimately worry about too much heating in today's atmosphere, the Earth would be too cold without the greenhouse effect of carbon dioxide, methane, water vapor, and nitrous oxide that keeps it warmer. Liquid water currently exists on Earth only because of these greenhouse gases, which absorb infrared light and warm the planet, thereby establishing equilibrium.

The *habitable zone* is the region where conditions are such that life can survive. It is the "Goldilocks" region that is just right to allow for stable liquid water. Too far away from the major heat source—the Sun—and water will be ice. Too close in and the water won't condense onto a planet's surface in the first place. Water might exist below the surface of a planet too, though that is unlikely to house the diversity of life that a large ocean can promote.

The outer habitable boundary with respect to water is sometimes defined as the distance from the Sun at which carbon dioxide would begin to condense out of the atmosphere, yielding a zone extending about one-third farther from the Sun than the Earth. It is sometimes alternatively defined as the region where enough carbon dioxide and water remain in the atmosphere to keep water from freezing, leading to a larger habitable zone around the Sun roughly two-thirds farther out than Earth. To put this in perspective, Venus falls into both categories, but Mars falls only into the second, and the outer planets, being too far away, fall into neither.

Even without knowing how it arose, we do know that water has been around almost from the planet's beginning. Yet the luminosity of the Sun has changed—having increased substantially since its formation—and the atmosphere has changed too. There is therefore a more limited region known as *the continuously habitable zone*, which is the region that could have supported liquid water over the lifetime of the planet. According to current climate models, the continuous habitable zone is a more restricted region within fifteen percent of the Earth-Sun distance. Of course that is defined by today. In another four billion years or so, the Sun will turn into a red giant, and a few billion years after that, it will burn out completely. According to current models, no forms of Earth-bound life—simple or complex—will survive in that distant future.

However, before we worry about that remote and dismal fate,

more pressing matters encroach. A critical one is the stability of the Earth's temperature and what it means for life as we know it. In our current society, relatively small temperature variations can have big effects on coastlines, agriculture, and human habitability. But for understanding the evolution of life, much more coarse-grained temperature considerations come into play. On Earth, carbon is essential, and atmospheric carbon must constantly be replenished.

On other planets, methane and carbon dioxide clouds might be relevant too. On this planet, the processes that regulate carbon in the atmosphere are critical. Carbon is removed from the atmosphere when it dissolves in rainwater or is absorbed through photosynthesis in plants, and it is replenished when recycled back into the atmosphere through plate tectonics and the constant weathering of rocks. Carbon returns when the ocean floor that is created mid-ocean later is lost in subduction zones where elements react to produce carbon dioxide that escapes through volcanoes, hot springs, and other vents. Carbon is also returned slowly through uplift and the creation of mountains, and it is recycled quickly through the burning of fossil fuels. All of these processes affect the supply of atmospheric carbon, which in turn is critical to regulating the Earth's temperature.

Long-term climate stability might have been another precursor to life's development. On Earth, this stability relied not only on oceans and internal heat sources to drive the plate tectonics that created a greenhouse layer, but also on stellar evolution, on a low impact rate of asteroids and comets, and on the presence of the Moon, which stabilizes the Earth's spin rotation axis. These conditions were probably most critical to the life that formed in the last five hundred million years, with its larger plants and animals, though some climate stability was probably important to early microbial life during the first three billion years too.

A stable astrosphere probably was very likely important to life's emergence. Too many cosmic rays hitting the planet—or too many asteroids or comets for that matter—and many types of life wouldn't have stood a chance of forming. Anything that did successfully emerge would probably have been quickly destroyed. A planet hosting life has to be far enough away from the Sun to avoid excessive solar radiation but perhaps should be sufficiently close to be protected by outer planets from asteroids. Whether or not this is necessary, Jupiter certainly does play the role of the Earth's big brother—or bouncer—protecting its smaller "sibling" from extraterrestrial attacks and making the development of life that much simpler.

Also protecting the planet is the stellar wind—discussed in Chapter 8 in the context of defining the edge of the Solar System—which interacts with interstellar material to create the *heliosphere*. The galactic cosmic ray rate is relatively low inside this region, possibly stabilizing the Earth's climate and protecting any emerging life from its more direct destructive influences.

Surprisingly, we currently live in a region—300 light-years across—called the *Local Bubble,* which is a vacuum-like domain with very low hydrogen density within the interstellar medium in the Orion Arm of the Milky Way. Only recently –perhaps in the last few million years—did we enter this warm, low-density, partially-ionized region, with its relatively sparse interstellar environment. During this time, the region enclosed by the heliosphere boundary—where the solar wind dominates over the interstellar medium—has been exceptionally large. We don't know if it is mere coincidence that the emergence of hominids on Earth coincides with a time interval when the Local Bubble's cavity surrounded Earth or if such an anomalously low gas and cosmic ray density was instrumental in thc formation of complex life.

METEOROIDS AND THE DEVELOPMENT OF LIFE

The meteoroid that created the Chicxulub crater certainly played a role in the later course of life's development by eliminating most existing species and paving the way for others. Though the numbers aren't very precise, it seems that most large meteoroids date to a time close to or coincident with a mass extinction. Iridium layers, microtektites, and shocked quartz near extinction boundaries lend support to a possible role for impacts that is potentially worthy of further investigation, as do actual craters whose timing seems to coincide with some landmark life-altering events.

Even so, many of the suggestions below are speculative. Despite the enthusiastic post-Alvarez wave of proposals for meteoroid triggers, asteroids and comets are certainly not the complete explanation for the destruction—or the origin—of life on the planet. The K-Pg event is the only reliably established impact-induced extinction. The evidence that climate change and large igneous eruptions played a role in extinctions at the end of the early Cambrian, the end-Permian, the end-Triassic, and the mid-Miocene is perhaps more convincing than some of the impact suggestions. So don't get overly excited by the speculations I'll now present. But evidence does point to an interconnected system. Since some of the larger impacts have occurred at times that approximate the age of the Earth, the origins of life, and the beginning of civilization, it is worthwhile investigating any possible connections to the extent that we can—even if the evidence is not overwhelming.

Of the five major mass extinctions, the end-Devonian—which occurred between 360 and 400 million years ago—is second only to the K-Pg extinction in terms of evidence for an extraterrestrial role. Several impacts probably occurred at this time, most likely triggered by an asteroid that fragmented or by a disturbance that triggered multiple comet impacts of the sort we soon will consider. Though precise timing measurements don't necessarily support a significant role for meteoroids in this extinction event—and the species loss in this time frame seems more the consequence of limited speciation than actual extinctions—it's interesting that in 1970, well before Alvarez's suggestion about the K-Pg extinction, the paleontologist Digby McLaren had suggested that an asteroid impact might have been responsible for this earlier event.

Most other suggestions for connections between impacts and extinctions concern more minor events, such as the regional extinction in North America 74 million years ago. Many crocodile species, some aquatic reptiles, some mammals, and several dinosaur species went extinct at this time, which seems to coincide with that when the

Manson impact structure in Iowa originated. The timing of the late Eocene events about 35 million years ago, which consisted of multiple extinctions in the sea and some reptile, amphibian, and mammal extinction on land, coincide approximately with some impacts too. Evidence includes the Popigai astrobleme in Russia, a recently discovered 90-kilometer wide astrobleme in the Chesapeake Bay near Washington DC, and another smaller one off of Atlantic City, New Jersey. The DC feature was cleverly discovered by the identification of a boulder field with an impact-induced tsunami deposit, which was followed up by seismic profiling and the examination of drilling cores. Higher-than-normal iridium levels and excessive interplanetary dust from this time suggest that a comet shower might have been the culprit responsible for the multiple impacts.

The late Eocene event also exhibits evidence for extraterrestrial interference that relies on a different method—geochemical evidence—which might ultimately help supplement the frustratingly sparse impact record. Ken Farley from the California Institute of Technology and collaborators have shown how to learn about impact events from an isotope of helium that traces interplanetary dust, a substance which can be enhanced during comet showers. Their very interesting result shows a rise in helium-3 between about a million years before the impacts that produced the Popigai and Chesapeake Bay craters 36 million years ago and a million and a half years after. The dust provides strong evidence for a comet shower, perhaps triggered by an impulsive perturbation of the Oort cloud, which is a topic we will return to in the next couple of chapters.

To complete our list of speculative impact proposals, a minor mass extinction in the late Miocene period around ten million years ago seems to coincide with an iridium anomaly and glass spherules. Interestingly, Farley also has identified an enhancement of helium-3 at this time. The timing and temporal evolution of the dust in this case is more consistent with asteroids—the known collision that produced the Veritas asteroid family in particular.

The role of impacts in life's creation is even less evident than their role in life's destruction. But some people also entertain the possibility that impacts played a role in this process too. I'll even mention the imaginative possibility that has been suggested that some of the more dramatic events in the Bible and mythology, or even unexplained prehistoric formations such as Stonehenge, might have been motivated by mysterious or mystical-seeming events with an extraterrestrial meteoroid-induced origin. Hewing closer to science, researchers suggest that early impact events might have blown away parts of the atmosphere and even the oceans, delaying or restricting life's progress on Earth. But these events might also have created environmental conditions that led to life—by creating hydrothermal systems that supported prebiotic chemical reactions, for example.

Charles Frankel in *The Extinction of the Dinosaurs* notes the coincidence of the introduction of complexity in the Precambrian era about two billion years ago and two known enormous impact craters from that time. Though not very convincing—oxygen's role was probably more critical—his point about the timing is intriguing. Another equally remote possibility is that impacts played a role in the much later Cambrian explosion (explosion here refers only to the escalation in diversification of life), 550 million years ago—presumably by eliminating many extant species and making room for new ones. Even without a known mechanism for forging a connection to life, evidence for impacts can be found in Australia and elsewhere. Australia's Lake Acraman Crater, more than 100 kilometers in diameter, is surrounded by an ejecta layer containing iridium and shocked quartz that extends 300 kilometers east into the Ediacaran fossils, whose formation immediately preceded the Cambrian explosion. Further evidence comes from the Yangtze Gorge in southwest China. Remarkably, immediately above the boundary layer with chemical evidence of an impact, trilobite fossils occur, indicating complex life began forming in the ocean immediately after whatever event deposited the more exotic elements.

Another speculation involves the surprising fossilized meteorites, shocked materials, and crater observations that strongly support a cluster of impacts in the Ordovician, with the peak rate in the mid-Ordovician, about 472 million years ago—precisely coinciding with a possible burgeoning of speciation in marine life in particular. The idea of fossilized meteorites is pretty impressive, so I mention this discovery here, even if the coincidence with life diversification is almost certainly too speculative to take seriously. The original clue for impacts at this time came from an isolated boulder found in 1952 in the sedimentary rock of Sweden, where it clearly didn't belong. But it took 25 years for it to be recognized it as a fossilized meteorite—a meteorite for which all the actual material had been replaced except for chromite, which is a form of rock that is highly resistant to weathering. Since then, almost a hundred such meteorites have been found in the vicinity, with the net amount of material pointing to a breakup of a 100-150 kilometer wide object half a billion years ago—leading to meteorite and micrometeorite dust that fell to Earth at an elevated rate for millions of years. The pieces might even form an asteroid belt that continues to slowly rain down objects today.

Some of the suggestions above for meteroids' role in either the destruction or creation of life have questionable merit. But I'll conclude this section with a reliably established role for meteoroids, which is as a significant fount of resources on the planet. Interestingly, the materials brought by meteoroids were important for societies even before the iron age—early humans used meteoritic iron to make tools, weapons, and cultural objects.

Such deposits have been critically important for the modern era too. A lot of the gold, tungsten, nickel, and other valuable elements in the Earth's crust are accessible because of extraterrestrial objects that pelted the Earth. Even though planets and asteroids were formed from the same stuff, Earth's gravity pulled heavier elements into its core and most of them won't flow back to the surface. These materials have been replenished primarily from extraterrestrial objects raining down. Perhaps a quarter of meteoroid impacts have led to potentially profitable deposits—at least half of which have already been exploited. So, even if meteoroids hitting the Earth weren't necessarily instrumental to life's creation, extraterrestrial objects that have made impact on the planet have undoubtedly helped pave our way of life.

14

WHAT GOES AROUND COMES AROUND

Early in the twentieth century, the physicist Lord Rutherford, best known for his landmark discovery of the atomic nucleus, famously pronounced, "All science is either physics or stamp collecting." Though arrogant and a tad obnoxious, this statement does contain a kernel of truth. Science isn't solely about listing phenomena, no matter how beautiful and remarkable they might be. It is about trying to understand them. Scientists can gather facts through impressive and perpetually advancing methods, as biologists do today, for example, when they use DNA sequencing and other techniques to facilitate rapid data accumulation. But the information becomes true science only when the data are more completely understood—ideally through a comprehensive theory with which to test hypotheses and make predictions.

We have now investigated what is out there in the Solar System, what has hit the Earth, and what is known about extinctions from the fossil record. A good deal of scientific inquiry has gone into extracting, understanding, and interpreting all of this data. But some big questions remain, such as "Which of these phenomena are connected?" and "If so, how?"

One of the most intriguing but also highly speculative suggested astrophysical connections is that objects from outer space hit the

Earth on a regular basis, yielding periodic impacts separated by an interval of between 30 and 35 million years. If real, periodicity would be a very important clue as to what precipitates the deviant trajectories that convert safely orbiting objects into potentially dangerous missiles hurtling toward Earth. Many suggestions for perturbations exist, but very few could possibly give rise to a periodicity that would stand a chance of matching the existing crater record.

I will shortly try to do justice to Rutherford's point of view by evaluating whether meteoritical (too bad the nicer sounding "meteorological" was usurped by weather) events exhibit enough of a periodic pattern to merit a scientific explanation. But I'll first note an unrelated but far better established such connection: that between cyclical movements of the Earth through the Solar System and periodic variations in the planet's climate. These temperature variations, known as Milankovitch cycles, occur on much shorter time scales than the ones I will soon consider. They are named after the Serbian geophysicist and astronomer Milutin Milanković, who developed his ideas while a prisoner during World War I.

Milanković investigated the effect on climate of the Earth's changing eccentricity, axial tilt, and precession. Based on these considerations, he and later scientists established the existence of both a 20,000-year and a 100,000-year approximate periodicity in temperature patterns, which they found reflected in global ice ages. When visiting Zumaia in the Basque country of Spain, my guide there pointed out the readily visible layered structure of the rock. These layers are a result of these same temperature variations, which cause sedimentation rates to change periodically over time as well.

Milankovitch cycles notwithstanding, the search for crater periodicity—which reflects a much bigger time scale—is necessarily a bold enterprise and I don't want to oversell it. Present-day evidence of events that occurred on Earth millions of years in the past is sparse and has many uncertainties, such as the precise time at which they occurred. Only in rare instances do these long-ago events leave any information at all and even less frequently do they leave enough of an impression to get a detailed understanding. Yet as long as hypotheses are consistent with existing data and have the potential to teach us something about the world, scientists can meaningfully explore them. Any curious person would want to know not just what happened but also what the underlying causes might be.

We will now consider proposed suggestions for large, regularly spaced impacts that occur on multimillion-year-time-scales in the hopes of connecting them not to the motion of the Earth through the Solar System, but to the Solar System's motion through the galaxy. In studying cratering data and trying to explain what has been observed, we aim to better understand the dynamics of the Solar System and the Universe, as well as their underlying connections. The most interesting suggestions are those that lead to predictions with which we can test hypotheses—however unlikely a skeptic might deem them to be. Although many of the ideas about periodicity are speculative, the goal of this chapter is to carefully explain what we accept and what we expect will require further study.

ESTABLISHING PERIODICITY

Matt Reece and I didn't immediately embark on our investigation of the possibility that dark matter can explain periodic phenomena in the Solar System. Before introducing our own ideas, we wanted to first ascertain that the evidence for periodicity was sufficiently strong to be worth investigating further. Another consideration that was important to us was whether our contributions might help guide future observations and analyses.

When starting out, we met in my very messy office and discussed the messy status of existing ideas—clarifying what was already understood and trying to determine how best to proceed. Our first order of business was to investigate the evidence for periodicity and establish whether it was reliable or if periodicity was merely an intriguing word that some scientists bandied about.

We read a lot of the previous research. But plowing through the papers and disentangling the claims and the truth was more challenging than you might imagine. One result followed the next—with some scientists finding evidence for periodicity in one set of papers and other scientists identifying mistakes or omissions of the previous authors in the following set. The debate raged on, with no real resolution. After we wrote our recent paper, the skeptics of the evidence for periodicity certainly made their thoughts known. However, we were in the fortunate position of having no axe to grind ourselves. We were just curious, and I think this lent us some helpful objectivity.

The requisite underlying statistical analysis is indeed tricky. The geological record is sparse and inevitably will contain large gaps. Due to the incompleteness of the data, the precise way in which a researcher evaluates the record can influence his or her result. It's tempting to view data as sacred matter on solid ground, but a lot of interpretation goes into determining how to present and evaluate measurements that are statistically poor.

The grouping of the data makes a difference, for example. When scientists view the data as a time series, they are faced with critical choices that can affect the conclusion, such as how many points they should use, and where exactly in the time interval any particular piece of data should be placed. They also need to evaluate the duration of events and understand the implications of their choices for the signal strength during intervals of enhanced activity.

Papers written in response to the ones that demonstrated periodicity also stress the many possible statistical mistakes that might have compromised the investigations. Coryn Bailer-Jones, from the Max Planck Institute for Astronomy, in Heidelberg, Germany, leads the charge. He provides many objections—including the ones mentioned above. He is also concerned about "confirmation bias"—the fact that people are more likely to notice or report on results they agree with. Bailer-Jones thinks that maybe the authors are trying too hard to get a fit because of the closeness of the period to the period suggested for extinctions or the motion of the Solar System that I will discuss in the next chapter. But though many of his other objections are valid, this closeness isn't necessarily a bad thing. A coincidence of numbers might be just that. Or it might be an indication of some underlying scientific connection that will lead to future understanding.

However, another common error that Bailer-Jones and others point out is that you can't simply compare a hypothesis to a single competing model and treat that one alternative suggestion as a substitute for all the remaining options. For example, people often ask which fits the data better: the hypothesis that meteoroids strike on a regular basis or the proposal that the strike probability is roughly constant over time. Even if the periodic model works better than the assumption of complete randomness, the data might conform better to a different model still—such as one where the probability of finding a crater decreases the older the meteoroid strike. In other words, the favored model doing better than the single alternative suggestion doesn't necessarily mean it is right. Fortunately, researchers can address this mistake by broadening the repertoire of models that they compare to. In the absence of a definitive difference in probabilities, it makes sense to try a variety of alternative models and test whether the periodic one does best.

Identifying a periodic signal presents still further obstacles. In 1988, the geologist Richard Grieve and collaborators pointed out that imprecise dating can wash out any signal of periodicity—whether or not the signal is real. In 1989, Julia Heisler, then a Princeton undergraduate, and Scott Tremaine, then a professor in Toronto working at the Canadian Institute for Theoretical Astrophysics and now head of the astrophysics group at Princeton's Institute for Advanced Study, further quantified this effect by asking how much uncertainty you can get away with while still reliably identifying a periodic phenomenon. In a paper published in 1989, Heisler and Tremaine argued that an uncertainty of 13 percent makes it impossible to get a better than 90 percent confidence that a periodicity in the data exists. If the uncertainty is raised to 23 percent, then the probability of detecting a periodic signal is down to about 55 percent. Such uncertainties don't make it impossible to reliably establish a periodic effect—but it does make it more challenging.

PERIODICITY IN EXTINCTION EVENTS

The focus of these particular cautionary papers was periodic effects in astrophysics, which will be the focus of my research I'll soon describe too. But the initial stimulus for investigating the time-dependence of craters came from studying a superficially very different topic—the apparent periodicity of extinction events. The Princeton geologists Alfred Fischer and Michael Arthur were the first to make the observation that life seemed to wax and wane on a regular basis. They concluded in 1977 that the fossil record seemed to accord with a period of 32 million years. David Raup and Jack Sepkoski of the University of Chicago published a far more influential paper in 1984, in which they presented their own search for periodicity in the record of extinctions. At first Raup and Sepkoski found a broad range of possible periods—somewhere between 27 and 35 million years—before redoing their analysis and revising their estimate to the 26-million-year period, to which most scientists studying the subject have since returned.

Any idea as provocative as this one is not likely to go unchecked, and later research did find supporting evidence—if perhaps with a slight variation in time scales. In 2005, using a recalibrated time scale but the same fossil record, two University of California, Berkeley physicists, Robert Rohde and Richard Muller, identified a different periodic signal of 62 million years. Subsequent results went back and forth but interestingly, periodic signals of 27 million years and 62 million years both survive. In one of the more recent and thorough analyses, Adrian Melott, an astronomy professor at the University of Kansas, and the paleobiologist Richard Bambach from the Smithsonian National Museum of Natural History in Washington, D.C., found that most extinctions occur within 3 million years of a 27-million-year periodic template and furthermore happen almost always during times of decreasing diversity in species over the 62-million-year time frame, indicating both time frames might indeed be relevant. All the caveats about periodicity still apply, but weak evidence in favor of periodicity persists.

However, even if the apparent regularities in the fossil record do turn out to be real, it won't change the fact that none of the authors explain why extinctions should be periodic. As we have seen, species can die out from a variety of causes. Climate change, volcanism, impacts, and plate tectonics all seem to have played a role. Meteoroids might influence some mass extinctions, and certainly one did precipitate the K-Pg extinction event. But any purported periodicity in extinctions is unlikely to be the result of a single root cause. Given the distinct physical causation mechanisms, one might at best expect a superposition of different periodic phenomena, which will look fairly random in the absence of a very complete record.

Any attempt to relate potentially periodic extinctions to physical processes that trigger them is bound to be even more speculative than attempts to understand periodicity in a particular physical phenomenon, such as extraterrestrial impacts, alone. Meteoroid hits are challenging enough to investigate. Coupling them with uncertainties about extinction events is bound to go down a convoluted rabbit hole of trouble.

Because of these uncertainties—apart from the lone well-established meteoroid/K-Pg connection—the rest of this book will shy away from further speculations about extinctions, intriguing as they may be. I will instead focus on the potential connection between periodic events in the cosmos and periodic impacts that are big enough to leave an imprint in the crater record. The study of impacts has the advantage that the crater record is directly related to astrophysics and—unlike potential causes of extinctions—doesn't suffer the intervening messiness of climate, environment, and biology.

Impacts offer a fascinating opportunity to explore a connection between phenomena on Earth and events in the Solar System as a whole—a unique lens through which to learn more about the cosmos. Random meteoroid strikes don't call for any particular explanation. Periodic meteoroid strikes very likely do. If meteoroid impacts truly occur on a regular basis, the time-dependence could signify an underlying cosmic cause.

Chapter 21 will explore what my collaborator and I determined will be a more reliable way of examining the data in the future and the slightly stronger support for periodicity that even existing data can yield. For now I will present some representative conclusions in the older literature, without going into detail about the precise statistical method or the choice of data set.

We'll now see that the older literature shows some evidence in support of periodicity, but the evidence is too weak for us to be confident in the result. Such ambiguous conclusions might disappear with better data and more careful analyses or they might end up proving more robust. For now, think of these results as an indication of the attention that scientists in the past have given to the search for a periodic component in the cratering data—and the perhaps optimistic conclusions they have found—rather than as a comprehensive survey or conclusion.

PERIODICITY IN THE CRATER RECORD

In any case, restrictions on the data are required when looking for crater periodicity. Analyses focus on larger and more recent craters. Anything that hit too long ago should leave a less reliable imprint than a similar but more recent one. On top of that, although the number of smaller craters is far greater than the number of bigger ones, searches for periodicity should include only larger craters. Small objects hit the Earth all the time but—aside from collisional cascades from the asteroid belt—most such events are random. The bulk of the objects that create small craters strike indiscriminately. As the next chapter explains, true periodicity seems possible only for comets, and, among these, only those from the distant Oort cloud.

So there is a trade-off between recording bigger numbers (which favors a smaller cutoff on size) and identifying a periodic phenomenon more reliably (which favors a bigger cutoff). The optimal choice is not known. The analyses in the literature all use different sizes for the boundary, which anyone has to bear in mind when evaluating the earlier research results. In the research that Matt and I did, we ultimately decided on craters larger than 20 kilometers that had arrived within the last 250 million years. Our time cut-off of 250 million years seemed big enough to allow for reasonable statistics but recent enough to be more reliable. Twenty kilometers seemed a good choice for the cut-off on size as it is big enough to require an impact from a kilometer-sized object but not so big as to preclude statistically relevant data.

Even with these constraints, reliably identifying periodicity in the crater record is a tall order. The imprints of craters that remain from the course of the Earth's history is incomplete—with only a small fraction of them still visible today. Moreover, the dating of the craters—even if and when discovered—isn't always sufficiently precise to reliably extract the time dependence of events. Complicating matters further was that researchers employed different data sets. Even with the same data, investigators sometimes used different time intervals or grouped data in different ways. The situation is further confused since, as discussed above, even if some impacts occur periodically, some strikes are nonetheless random. This means that we can at best expect a periodic component superimposed on a random one, which further compromises an already poor statistical record.

Nonetheless, motivated in part by Alvarez's 1980 proposal of a meteoroid-induced K-Pg extinction, as well as by the evidence for periodicity in extinctions, scientists forged ahead and searched for evidence of periodic impacts. In 1984, Alvarez and his UC Berkeley colleague—the physicist Richard Muller—started the ball rolling when they proposed a 28.4-million-year periodicity in craters with radius greater than five kilometers that were formed within the last 250 million years. Their result was based on a sample of only 11 craters and didn't rigorously account for uncertainties in the data, but many more comprehensive analyses soon followed.

Later that same year, the New York University biologist Michael Rampino collaborated with Richard Stothers from the NASA Goddard Institute for Space Studies to study a sample of 41 craters from between 250 and one million years ago and identified a 31-million-year period in extraterrestrial impacts. In 1996, scientists in Japan suggested something similar—a 30-million-year period using craters from the last 300 million years. In 2004, Shin Yabushita, an applied mathematician from Kyoto University and one of the authors of that research, did a more subtle analysis with craters from the last 400 million years in which he weighed the importance of each crater differently according to its size. He thereby derived a 37.5-million-year period from a set of 91 craters. These analyses all found some evidence for regularity in the crater record. But the periods identified didn't agree well enough to robustly support the results.

In 2005, William Napier, a professor at the Buckingham Centre for Astrobiology, in England, did an interesting study in which he claimed that impacts tend to occur in groups separated by about 25–30 million years, with each episode lasting about one to two million years. His sample of 40 craters included those that were bigger than three kilometers from the past 250 million years. He found the biggest hits occurred during relatively short intervals and noted that the K-Pg extinction was among them. The evidence for periodicity was weak, however, and he deduced a range of scales—depending on how the data was interpreted—in which 25 million years and 35 million years seemed to dominate.

Napier himself recognized his evidence was insufficient to make a strong case, and even pointed out that with so much more data than Alvarez originally had, one might have expected either a stronger signal or for the signal to have completely disappeared. He suggested that a plausible explanation for the ambiguity of his result was that there is a relatively constant ratio of random and periodic events so that a signal wouldn't cleanly emerge—even after a tripling of the data set.

Napier also made some intriguing suggestions concerning comets versus asteroids as a potential source for his admittedly weak signal. Though he thought that the smaller meteoroids that he excluded from his analysis probably originated in the asteroid belt, he suspected that comets—not asteroids—were primarily responsible for the larger ones he found. He reasoned that the supply of large asteroids was inadequate to explain the bombardment episodes' needed intensity—arguing that too many large asteroids would have had to break up on too short a time scale to account for what was observed. Napier pointed out that the inadequacy of the crater record actually bolstered his case. If most craters don't survive, the number of hits had to have been even greater than the number he could identify from astroblemes on Earth. If we know of a few large craters from a single bombardment episode, very likely many more hits occurred for which there is no longer evidence.

Napier reasoned further that fewer than 1/25 of the asteroids that are perturbed into orbits that cross that of the Earth actually hit the planet. Most are sent out of the Solar System or fall into the Sun. Accounting for both effects, Napier concluded that hundreds of asteroids would have had to have been injected into near-Earth orbit from the breakup of a parent asteroid of at least 20–30 kilometers across to account for his data. This breakup would have to have been due to collisions. But collisions break up big asteroids far too infrequently to explain such numbers. Since neither the short impact time of one to two million years nor the time frame seemed appropriate for an asteroid-based explanation, he suggested that comets are the much more likely source of the periodic bursts he identified. Although his conclusions are by no means proven and we now know that some asteroids take a "fast track" of one to two million years, they do suggest the possible relative importance of comets over asteroids for some significant impacts and perhaps even ways to ultimately distinguish them.

THE "LOOK-ELSEWHERE EFFECT"

These are all tantalizing observations. However, none of the results presented above have the statistical significance required to definitively establish a periodic effect. But one additional tricky issue arises when analyzing statistical significance, and this one, which probably accounts for the bulk of why the literature presented contradictory results, is surmountable.

You might think if you make a hypothesis that data is periodic, you can simply try to match the data to a periodic function and evaluate how well the best-fitting periodic function does in explaining observations. However, this would yield an overly optimistic estimate. When you are not testing a single hypothesis but have many possible guesses—in this case functions with different periods—given enough possibilities, almost certainly one will prove to be a better fit to the data than random hits. But that doesn't make it right.

This subtle but somewhat obvious (in retrospect at least) problem is known to the particle physics community as the *look-elsewhere effect*. This phenomenon was the subject of a lot of discussion around the time of the Higgs boson discovery at the Large Hadron Collider (LHC)—the giant particle accelerator at CERN near Geneva that collides very energetic protons to try to produce new particles that can give insights into underlying physical theories. Though not the subject of this book, the results of the Higgs search illuminate an issue that scientists looking for periodicity also face.

The way experimenters search for a Higgs particle is to look for evidence in the data of the particles that the Higgs boson decays into and then measure how often they are found. Because most of the time when particles collide, a Higgs boson is not produced, the indication of the presence of a Higgs appears in the data as an elevated signal over a smooth *background* curve representing those events that occur even in the absence of a Higgs particle. If plotted appropriately, this elevation should occur at the correct Higgs mass value. So when experimenters present their data, they focus on "bumps," regions in the data where something—hopefully a Higgs boson—gives a sizable contribution over background.

The caveat is that statistical flukes (known technically as fluctuations) give rise to ups and downs in the data all the time. Sometimes a big fluctuation occurs. Though any particular fluctuation is unlikely, even an unlikely fluctuation should occur somewhere if you study a big enough range of masses. That unlikely event would have the appearance of a Higgs boson. But it would just be an improbable accumulation of background events at a given apparent mass.

When the experimenters first began searching, they didn't yet know the Higgs boson's mass.* They could measure this mass if

* Technically there were constraints due to precision measurements of other processes but generally these were neglected in the presentations based purely on direct searches for the Higgs boson itself.

and when they found the appropriate evidence, since the energy and mass from the outgoing decay products would be related in a way that would determine its value. But the researchers could determine the mass only after they saw a bump—not the other way around.

When the experimenters presented their data and discussed how likely or unlikely any bump they identified would be in the presence or absence of a Higgs boson, they had to account for the uncertainty in value of the mass of the Higgs boson. Because statistical fluctuations could occur anywhere, and any one of those might also be interpreted as a Higgs boson decay, the statistical significance of any particular bump was compromised by the greater likelihood that some fluctuation would occur somewhere. The experimenters were aware of this, so they presented the significance of their result taking into account this look-elsewhere effect. The look-elsewhere effect tells you the result is much more significant if you know the Higgs boson mass in advance. If you don't, a bump is more likely to be a fluctuation since you are multiplying the likelihood of an anomalously large rise in data by the number of places this unlikely event could have occurred. Only after the experiments had generated enough detectable Higgs bosons to present a statistically significant result—even with the look-elsewhere effect taken into account—could physicists finally claim discovery.

Similar considerations apply when looking for periodicity in the crater record if you don't have advanced knowledge of what period you are searching for, though astrophysicists use a different name: *the trials factor.* If you allow for enough different periods, one of them is likely to look better than none at all—namely totally random data. As it turns out, models that incorporated periodic meteoroid strikes matched the data well—at least better than a model that assumed the hits were completely random. But since no one knew what period to expect, any statistical significance a researcher could surmise based on a single better fit was lower than he or she would naively conclude. Given enough possibilities, each with its own possible statistical uncertainty, eventually some periodic function is bound to look like a reasonably good match to the data.

This goes a long way toward explaining the discrepancy between Coryn Bailer-Jones's results, where he doesn't find statistical evidence for periodicity, and those of his colleagues who do. Both did their respective analyses correctly but Bailer-Jones factored in that we don't know the period in advance. Without additional input, a signal needs to be strong enough to overwhelm this compromising effect. And it looked at first as if the signal wasn't sufficiently robust.

The good news is that we do have additional input, and we can take it into account. We know what the galaxy is made of since, to some extent, astronomers have measured its content and its gravitational pull. If periodic effects are precipitated by the motion of the Solar System, we can put together all that we know about the galaxy and the Sun's location in it to predict its motion and compare the prediction to the data. When applying a triggering mechanism that I'll introduce in the following chapter, this is precisely what Matt and I decided to do.

15

FLINGING COMETS FROM THE OORT CLOUD

You might have seen the coordinated dances performed by the Rockettes at New York City's Radio City Music Hall or by ensembles on some old TV shows, in which a large number of beautifully dressed women synchronize graceful motions around a circle. Some of the formations involve spokes of dancers emanating from a common center, while others are composed of individuals forming rings—one inside the other. The performers seamlessly keep their circles intact, making it easy to forget how difficult the precise relationships among the individuals are to maintain. This applies especially to the outer members, who need to move more quickly and are also more distant from the inner region from which the coordination and directions emerge. You might occasionally see a dancer in the farthermost ring, who faces these greater challenges, mess up and get out of sync. But so long as she doesn't fall, it's not such a big deal. Though the error will take away from the performance's beauty and perfection, which lies in the dancers' synchronicity, nothing dramatic or disastrous ensues.

Icy bodies in the Oort cloud—tens of thousands of times farther away than the Earth is from the Sun—face challenges akin to dancers in the outer ring. Its members are so remote from the Sun's gravitational pull that they are in a relatively precarious equilibrium.

A sufficiently strong disturbance can cause a body, like the less precise outer dancer, to slowly move out of its expected position. If an Oort cloud object comes too close to the inner solar region, enough nudges—or one single more dramatic push—will send it out of its orbit altogether. When this happens, that body will deviate far more from its path than the errant dancer, running the risk of hurtling toward the inner Solar System and possibly even toward the Earth.

Near-Earth asteroids as well as some errant short-period comets can also be jostled by planets or other local objects so that they occasionally hit the Earth. But such impacts are almost certainly random. Mechanisms for triggering periodic disturbances have been proposed only for comets from the Oort cloud. The Oort cloud, the lone source of long-period comets entering the Solar System and also probably the source of most comets coming near the Sun, is also the only suggested source of regularly spaced comet strikes. The suggested periodicities in the extinction and crater records that we considered in the previous chapter have made for a good deal of interest in identifying what might trigger disturbances that could regularly send the Oort cloud's icy bodies into the inner Solar System.

In this chapter, I'll first briefly address the issue of whether comets or asteroids are more likely to have created big impacts. I'll then review some of the original proposals for what might dislodge objects from the Oort cloud to create comets that can impact the Earth. Although these older ideas failed to account for the suggested regularity, they are nonetheless interesting in that they encouraged new ways of thinking about galaxy interactions. They also paved the way for our later, more promising proposal based on our newly suggested dark matter idea.

ASTEROIDS VERSUS COMETS

If the impactor responsible for Chicxulub was caused by an asteroid, dark matter had nothing to do with it. But if it were a comet that wreaked the devastation, an exotic dark matter trigger just might have been the culprit. In his book *T. rex and the Crater of Doom,* Walter Alvarez used "comet" as his default when discussing the impactor responsible for the K-Pg extinction, with the understanding that no one could definitively establish whether a comet or an asteroid had been responsible. Distinguishing the effects of comets and asteroids that have created craters—particularly those that fell to Earth millions of years ago—is difficult. If no one observed the trajectory, we usually have no way of knowing whether an object that hit was a comet or an asteroid. For the comet that destroyed the dinosaurs, the jury is still out.

We do know that comets and their fragments hit the Earth less often. Estimates of the relative frequency of comet impacts compared to those of asteroids range between 2 and 25 percent. This small rate corresponds to the low number of near-Earth comets. Of the more than 10,000 near-Earth objects that are now known, only about 100 of these are known to be comets, with the rest consisting of asteroids or smaller meteoroids.

But larger impacts don't necessarily arise solely from objects that are already nearby. Distant comets can sometimes escape their orbits to occasionally hit the Earth too. An intriguing study by the extraordinary astronomer Gene Shoemaker argued that although asteroids predominate in smaller impacts, comets might be more important for larger ones. Shoemaker plotted the number of impacts versus size, and found that the results seemed to depend on two separate populations. The smaller impacts all fell along a nice curve, but there were many more large impacts than this simple curve could accommodate. Knowing that asteroids produced the smaller impacts, Shoemaker hypothesized that a new source of impacts must have produced the larger ones—arguing that what he was witnessing was the sum of two separate curves representing two independent contributions. His guess was that the source of the larger impacts was comets.

Comets have the further feature that they carry a disproportionate amount of energy compared to asteroids, since they are generally moving at faster speeds—up to 70 or more km/sec as opposed to 10 to 30 km/sec for asteroids. Typically, a ballistic missile travels at less than 11 km/sec, an asteroid at about 20 km/sec, a short-period comet at more like 35 km/sec, and a long-period comet at 55 km/sec, though faster speeds occur too. (See Figure 32.) Kinetic energy grows not only with mass, but also with the square of the velocity. Comets' greater speeds mean that even less frequent comet impacts, or ones from smaller objects, could in principle do greater damage than slower-moving asteroids.

Shoemaker furthermore did chemical analyses that supported the comet proposal—though in fairness it should be noted that scientists doing such analyses have argued both ways. In favor of the competing asteroid hypothesis are isotope ratios and surviving meteorite fragments matching those of *chondritic* asteroids, which contain millimeter-size spherical pieces that were once molten droplets created in nebular storms 4.56 billion years ago, during the Solar System's formation. But the evidence is not yet decisive. We don't know the isotope ratios in comets, so we might find that they are similar too. Furthermore, more recent research argues for a lower iridium and osmium level than had formerly been believed, which would be more consistent with a comet interpretation.

In 1990, the astrophysicists Kevin Zahnle and David Grinspoon argued for a comet impact at Chicxulub using very different reasoning. They proposed that comet dust entered the Earth before and

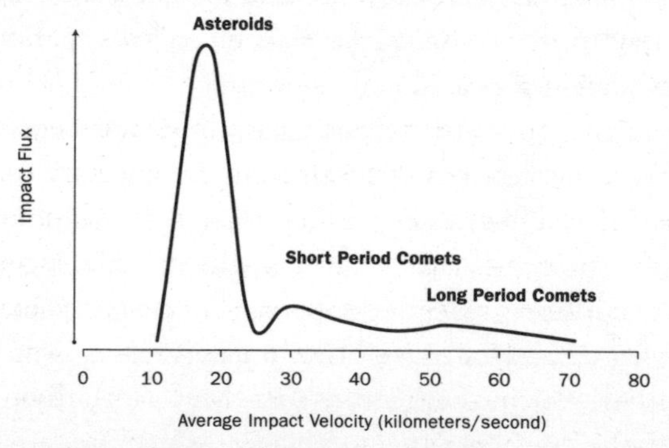

[FIGURE 32] Average velocity of impacts on the Earth from asteroids, short-period comets, and long-period comets in kilometers/second. The curve also illustrates the expected relative fluxes of the three types of objects.

after the K-Pg extinction event in order to explain the amino acids found in the sediments surrounding the K-Pg layer. Since dust particles get suspended in the atmosphere and fall slowly and thus would reach the ground intact, the dust could in principle be the result of a comet that disintegrated over a long period of time—raining material down onto the planet.

One reason that comet strikes might happen more frequently than would otherwise be expected is that when Jupiter swings comets around, it sometimes breaks them up into fragments. If and when this happens, the likelihood of connecting with Earth increases since several of these fragments might then cross the Earth's orbit. Some astronomers speculate that this occurred as recently as a few thousand years ago and cite excessive comet dust in the inner Solar System as evidence.

The relatively fresh Shoemaker-Levy comet hit on Jupiter was a spectacular illustration of the destruction such comet fragments can do. Carolyn Shoemaker first spotted the comet near Jupiter in 1993, and followed it along with her husband, Gene, and another colleague, David Levy. They noticed that the comet had an unusual appearance, appearing not as a single streak in the sky but as an arc punctuated by spherical bright spots. Soon afterward, through a more precise observation, the astronomers Jane Luu and David Jewitt were able to identify at least 17 separate pieces that spanned an arc resembling a string of pearls.

Astronomer Brian Marsden from the aptly named Central Bureau for Astronomical Telegrams deduced from its trajectory that its unusual structure was the result of a too-close flyby of Jupiter, whose gravity broke the comet into smaller fragments. He suggested the possibility of a future Jupiter close approach or even impact. Astronomers followed up and calculated that Jupiter's gravity would indeed trap the pieces, which would return for a head-on collision between July 16 and July 22, 1994.

Sure enough, right on schedule, the first fragment dove into Jupiter's atmosphere with a speed in excess of 60 km/sec. The region that was visibly affected was at least the size of the Earth. The atmosphere was illuminated by dust that preceded the actual fragments, which themselves created a brilliant flash. This generated effects similar to those surrounding Chicxulub—but this time the damage occurred on Jupiter. Since the fragments were less than 300 meters across and the initial comet that created the fragments was at most a few kilometers big, the energy released was far less than that from the object that created Chicxulub. Nonetheless, it was an impressive sight.

Impact craters on the Jovian moons indicate this was not the first time that such dramatic capture breakups and impacts have occurred in the region. And, if the periodic meteoroid idea turns out to be correct, it will be further evidence that comets have been important throughout the Solar System's existence. The association of such astrophysical phenomena with planetary surfaces reminds us that even seemingly abstract theoretical research might ultimately help explain our own existence.

TRIGGERS

Although no one can be certain, I will assume for the rest of this book that comets from the Oort cloud are responsible for big impacts. It is the only known possibility for which we can potentially explain periodic hits. Although a perturbation of an icy body from the outer Solar System that sends it toward our planet's path might sound like science fiction—not mistakenly so, since it often is—this sequence of events is also science.

Recall that the furthest reaches of the Solar System contain the Oort cloud—a hypothesized somewhat spherical collection of minor bodies that might extend beyond 50,000 times the Earth–Sun distance. The evidence for the existence of this huge source of comets—too far away to directly observe—is precisely the visible comets that have entered the inner Solar System.

In contrast to the situation for the dancers mentioned earlier, the pull of the Sun—and not the mutual interaction among Oort cloud objects—is responsible for keeping Oort cloud icy bodies in their orbits. But the Sun only weakly gravitationally binds the objects into the cloud, which is so enormously far away. Gravity's strength decreases according to the inverse square of the distance, so its influence on an object tens of thousands of times further away is less than 100 million times less strong. The Sun's pull on a comet in the Oort cloud is that much less strong than the Sun's pull on the Earth. In such a loosely bound environment, even relatively small disturbances can potentially alter an Oort cloud object's trajectory, ultimately kicking it out of its orbit—dispatching it out of the Solar System altogether or sending it hurtling down a path inward toward the Sun.

Though the astronomer Jan Oort later put the idea on firmer footing, the Estonian astrophysicist Ernst Julius Öpik proposed in 1932 that perturbations to comets at the outer edge of the Solar System (in what is now known as the Oort cloud and occasionally the Öpik-Oort cloud) sometimes send those icy bodies toward the inner region of the Solar System. Öpik had the whole story essentially correct—reasoning that some icy bodies would eventually become unstable and vulnerable to perturbations so that external influences would sometimes nudge them out of their orbits and into a path that was heading toward the Earth. He even suggested this could influence life here, though he didn't necessarily envision global devastation of the sort accompanying the K-Pg extinction.

Öpik's impressive work nonetheless left open the question of why the orbits became unstable or what the trigger was that precipitated their escape. These questions wouldn't be addressed until many years later, when the Alvarez proposal (and the Cold War with its images of massive devastation) entered the public consciousness and resuscitated interest.

Examples of objects that astronomers have suggested as perturbations include nearby passing stars and *giant molecular clouds*—enormous concentrations of molecular gas with mass between 1,000 and 10,000,000 times that of the Sun. But although the former jostles the orbits and the latter has some effect, neither is the dominant mechanism for sending comets en route to the inner Solar System. A nudge's influence depends on the magnitude and frequency with which it occurs as well as the density and mass of the icy bodies it acts on. Neither stars nor molecular clouds have both sufficient force and frequency to explain all the comets that we see.

In 1989, Julia Heisler and Scott Tremaine investigated a far more significant influence, which is the tidal force of the Milky Way. The Moon creates the familiar ocean tides through its distorting gravitational influence, causing the ocean to rise and fall by pulling differently on farther or closer regions of the Earth. In a similar manner, the galactic tide caused by the Milky Way bends the orbits of outer Solar System objects. The gravitational pull of the Milky Way acts differently on objects that are not in precisely the same location, deforming the otherwise spherical Oort cloud so that it is elongated toward the Sun and compressed along the other two directions.

Over time, the gravitational force from the Milky Way will tweak the paths of minor bodies into very elongated, or *eccentric,* orbits. Once sufficiently eccentric, the *perihelion*—the distance of closest approach to the Sun—will be so small that objects can be more readily injected into the inner Solar System. The tidal force at this point can be sufficient to send icy bodies out of the Oort cloud to increase the comet flux in the interior. The result is a slow and steady flux of comets reaching the Earth.

To make matters more interesting, it turns out that the dominant mechanism for dislodging icy bodies to send comets into the inner Solar System is not solely dependent on the tides, but rather involves both stellar and tidal perturbations working in concert. Though stellar perturbations are not ultimately the ones that usually create the comet showers, since they occur over much larger time scales than the tidal influence, they are essential for prepping the Oort cloud to a point where a tidal interaction can be pivotal. It's like a bicycling team in the Tour de France. The rest of the team helps the lead rider position himself so that he can make the final end run that earns him his yellow jersey. Because he crosses the finish line first, we generally only know the name of the winner—not the supporting *domestiques*. Even so, the other riders played an important role. Similarly, although the immediate trigger for dislodging the comets is the tidal force, the reason it can exert sufficient pull is that stellar perturbations have already jostled the orbits sufficiently that some are in precarious positions where only a relatively minor nudge will send the comet into the inner Solar System. Stellar encounters are essential, but the actual trigger for the comets—the one that gets the credit—is primarily the tidal force.

The distance at which the galactic tide dominates over the Sun's gravitational inward pull is about 100,000–200,000 AU from the Sun. At the outer boundary of the Oort cloud, the Sun's gravitational influence no longer suffices to maintain stable orbits. We have just seen how, farther in, tidal effects perturb borderline stable orbits, occasionally dislodging a minor Solar System body and sending it to the inner Solar System. Closer in still—in regions that have been accessible to observations—tidal effects pale in comparison to the Sun's gravitational pull. So it is only in the Oort cloud that tidal effects can jostle weakly bound comets significantly. And in all likelihood these tidal effects are responsible for 90 percent of the comets that originate there.

So the Milky Way contains the means to disturb comets to send them into paths toward the inner Solar System through a gravitational influence that physicists and astronomers now understand. But this mechanism—although important and interesting—doesn't suffice to account for all comet showers or periodicity in comet strikes. In the absence of additional complications, the tidal force I've described leads only to a slow but steady stream.

Astronomers trying to explain periodic enhancements therefore made further speculative attempts to explain why the triggers for these comets might not be completely random, but would instead occur at regular intervals in the range of tens of millions of years. I will say up front that the proposed explanations I'm about to present did not succeed. But understanding these suggestions and why they failed helps guide the search for alternatives. One of these suggestions was the precursor to the dark disk proposal that I'll later describe.

THE NEMESIS PROPOSAL

The first—and the most colorful-sounding—suggestion for explaining periodic impacts was that the Sun had a companion star playfully termed Nemesis, and that Nemesis and the Sun orbited in a big binary system. Astronomers proposed a very elliptical orbit for this hypothetical companion to the Sun that would allow it to pass within about 30,000 AU of us every 26 million years. This 1984 proposal was an attempt to account for Raup and Sepkoski's suggested extinction periodicity by Nemesis's enhanced gravitational force on the Sun every 26 million years when it was closest. The suggestion was that at those times, Nemesis' gravitational influence would dislodge from the Oort cloud minor Solar System bodies that could then bombard the Earth as comets.

The roughly 30-million-year period for the enhanced encounters (and hence a heightened rate of comet strikes) requires a very big system, with a *semi-major axis* (half the length of the ellipse) of order one or two light-years. One problem with this proposal is that stars or interstellar clouds would make such an enormous binary system unstable, destroying the regularity of the presumed encounters and causing the rate to vary over the last 250 million years. This variation hasn't been seen.

But the real nail in the coffin for this idea is the much-improved infrared survey catalog of objects in the entire sky—which would now include Nemesis had it existed. Although measurements were inadequate in 1984 to decisively rule on the suggested object's existence, observations have dramatically improved since that time. NASA's Wide-Field Infrared Survey Explorer, which was launched in 2009 and collected relevant data until February 2011, should have seen this proposed red dwarf–type star had it existed—but it didn't. Having not seen a proposed Jupiter-sized gas giant planet either, the infrared observations also ruled out another similar proposed explanation based on a hypothetical new planet, which the idea's originator named Planet X.

PROPOSED TRIGGERS FROM GALACTIC MOTION

In light of these failed ideas, a few very different proposals based on the Solar System's motion through the galaxy's known components seemed like promising alternatives. These proposals didn't introduce anything new and exotic, but instead suggested that existing density variations that the Solar System encounters when passing through the galactic spiral arms or in crossing the galactic plane could induce variations in the Oort cloud perturbation rate. These repeated passages through high-density regions could in principle account for periodic comet showers.

Recall that the Milky Way is a disk galaxy, meaning most of the stars and gas lie in a thin disk, about 130,000 light-years across but only roughly 2,000 light-years in thickness. The Sun is located at a distance of about 27,000 light-years from the galactic center, and happens at this moment to be close to the galactic midplane—less than 100 light-years away. It is also at the edge of a spiral arm.

The spiral arms of the Milky Way extend from the galactic center in the radial direction as they wind around. (See Figure 33.) These spiral arms contain more gas and dust than the regions in between, and consequently are areas where young stars are more likely to form. They are also the site of an enhanced concentration of giant molecular clouds—the enormous concentrations of molecular gas mentioned earlier. When the Sun crosses these denser regions, the molecular clouds exert a greater gravitational force that could in principle cause more extensive perturbations and thereby generate a periodic enhancement in impacts.

One potential problem with this proposal is that the spiral arms don't exhibit perfect symmetry or even have a fixed rotation rate relative to the Sun. Therefore, the Sun probably doesn't cross them at a precisely periodic rate. However, since the structure, kinematics, and evolution of the spiral arms are currently only poorly understood, any conclusion ruling out the spiral arm option on this basis alone might turn out to be premature. In any event, until periodicity is better

[Figure 33] The spiral arms of the Milky Way with the location of the Sun (size not to scale) indicated.

established, this lack of perfect regularity in the predictions doesn't necessarily rule out a match to the data, which might turn out to exhibit only approximate periodicity too.

However, two other factors make spiral arms a poor explanation for any observed enhancement in the impact rate. The first is that the average density of gas in the spiral arms is not elevated enough to account for periodic impact enhancements. If the density doesn't change by enough, any enhancement during spiral arm crossings will be too small to register.

The further issue is that the Solar System doesn't cross the galaxy's spiral arms all that often. There are only four big spiral arms and maybe two smaller ones and a galactic year is pretty long, so there have been fewer than four crossings of the larger spiral arms in the last 250 million years. In fact, because the arms move in the same direction as the Solar System (though at different speeds), the crossings are probably 80–150 million years apart—far too rare to explain the record of extinctions or impact craters.

However, the failure of spiral arms to explain the period and periodic enhancements doesn't rule out vertical variations in density as a potential impact trigger and this proposal might well prove to be the more promising suggestion. Superimposed on its circular motion, the Solar System oscillates in the vertical direction in which it covers a much smaller distance (compared to the 26,000 light-year radius where the Sun is located along the plane), as is illustrated in Figure 34. As it orbits the galaxy in a roughly circular orbit, taking about 240 million years to complete the circuit in what is known as a galactic year, the Sun also bobs slightly up and down. This much smaller oscillation amplitude in the vertical direction of the Sun depends on the matter distribution in the disk, but a reasonable estimate is roughly 200 light-years—though we are currently much closer to the midplane than the maximum height, perhaps 65 light-years away.

This oscillatory vertical motion of the Solar System could potentially account for variations in tidal effects over time and thereby explain any periodic effects on the appropriate time scale. Because the concentration of stars and gas changes as the Solar System moves in and out of the somewhat denser region of the galactic midplane, the Solar System encounters different environments as it oscillates across it. If the density were to increase dramatically as the Solar System crossed the plane, the perturbations would as well, and consequently the rate of comets hitting the Earth could be enhanced at those times. Since galactic tides are the dominant perturber of the Oort cloud, density variations in the vertical direction within the ga-

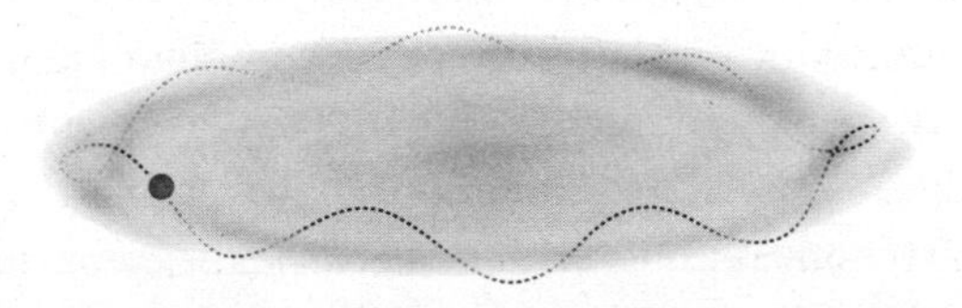

[FIGURE 34] The Sun oscillates up and down across the Milky Way plane as it orbits around the galaxy. During galactic plane crossing it encounters greater gravitational tidal forces. Note that the oscillation period was shortened in the picture for clarity and the Sun would make only three to four oscillations when going around.

lactic plane could potentially have sufficiently strong influence. The New York University professors Michael Rampino and Bruce Haggerty, who made this suggestion, gave it a colorful name too—the Shiva hypothesis—after the Hindu god of destruction and renewal.

Two features of the matter distribution in the galaxy are required for this scenario to match observations. First, the midplane density has to provide a gravitational potential that accounts for the correct oscillation period in the vertical direction. This condition is independent of any precise perturbation mechanism. If the Solar System doesn't cross the midplane at the correct rate, any enhancements at those times won't match the data.

The second feature is the one necessary to achieve the change in rate that could account for periodic comet showers—namely a sufficiently marked variation in density that would lead to a time-dependent influence on the Oort cloud as it passes through the galactic plane. These two features are relevant to any proposal of density enhancement at the galactic midplane. They rule out the proposals discussed here, and—as I will later explain—account for why the disk of dark matter, denser and thinner than the usual ordinary matter disk, might be a suitable alternative.

However, in 1984, Rampino and Stothers, relying on a more standard Milky Way composition, tried to explain the requisite density variations with giant molecular clouds—which are densest near the galactic midplane. Their reasoning was similar to that used in the spiral arm crossing suggestion—the matter concentration increases when the Solar System goes through the clouds. This proposal was quashed the following year when astronomers showed that the cloud layer is too big—it extends almost as far as the Sun's vertical oscillation amplitude, so the variation along the Sun's trajectory would be too small to register. Without additional matter, the encounters with molecular clouds are in any case too infrequent to account for an approximately 30 million year periodicity.

An alternative possibility was explored by Julia Heisler and Scott Tremaine—this time working together with the astrophysicist Charles Alcock. Having established the significance of the tidal influence from the Milky Way, they pointed out that although this effect alone predicted a fairly uniform comet rate, a kick from a nearby star did have the potential to create a comet shower. The question then becomes how frequently such encounters might occur and with what impact. How much of a variation in the rate of comets hitting the Earth should we expect?

The team estimated the expected rate by asking how often a star of a solar mass (the minimum mass needed to have the necessary impact when moving about 40 km/sec) comes within about 25,000 AU of an Oort cloud object (the minimum distance required to perturb it, since it is comparable to the distance of the Oort cloud from the Sun). It turned out one such encounter is expected every 70 million years. That isn't often enough to account for the suggested periodicity, but it could in principle account for a few such events within the last 250 million years.

Heisler and collaborators subsequently did a more extensive numerical simulation to make better predictions—taking into account the extra push the tidal force could provide. They found that stars had to be a bit closer than the Sun than they had previously thought. The real rate of predicted showers is therefore even smaller—more like once every 100 million or even 150 million years—far too infrequent to account for any periodicity that might have been observed. Subsequent, more detailed numerical analysis found the role of stellar encounters for triggering impacts was greater than they had found, but it was still insufficient to explain the data.

The conclusion of all this research is that without any new ingredients, the Solar System's gravitational potential doesn't change dramatically enough over a short enough time frame to yield an observable difference in meteoroid strikes in which the rate at regular intervals would have demonstrated an observable spike that dominated over the background rate. Although the Solar System crosses the galactic midplane on a periodic basis, comet showering due to a conventional matter distribution is not particularly elevated at those times.

So in broad overview, the situation turns out to be reminiscent of that with the spiral arms. The period predicted is too small and the change in density not sufficiently pronounced to give rise to a measurable periodic cratering of the sort the proposers had hoped to explain. Initial density measurements had suggested otherwise, but later calculations taking into account more recent data about the galaxy showed that suggestions that predated our work didn't generate the right frequency or correct episodic enhancement to match the crater record. The too-long prediction for the period rules out all the galactic plane proposals unless some new and as-yet-undetected component of matter is present in the disk.

Putting together the best available measurements—which, like the evidence for periodicity itself, changed a lot over time—Matt Reece and I ultimately concluded that without a component of so-far undetected matter in the disk, the up-and-down oscillation period was too long to account for the data suggesting periodicity. Not only was the distribution too smooth to generate an impulsive change in the cratering rate, but the familiar Milky Way disk, if composed only of normal matter, is too diffuse to give rise to the correct periodicity.

Though not in themselves sufficient to explain any potential periodicity, the above proposals nonetheless taught me and Matt the fundamentals we needed in order to proceed. We learned that tidal effects create sufficient perturbations to drive comets into the inner Solar System during and near disk crossing. But we also learned that known astrophysical sources do not create the hoped-for periodic effects. None generate a sufficiently abrupt tidal influence to account for an enhancement in comets reaching the Earth.

This left us with two possibilities. Perhaps the more likely one is that the observed periodicity isn't a real effect. The evidence is not all that strong in the first place and many accidents can conspire to give the appearance of a periodic effect. The second, more speculative, but far more interesting, option is that the structure of the galaxy is different from what is commonly assumed, in which case the tidal effect could be bigger and more dramatically varying than anticipated. This was the path we decided to explore. And it paid off.

As the next part of the book explains, when Matt Reece and I accounted for what is known about the density of ordinary matter in the Milky Way plane and the measured position and velocity of the Sun, we found that agreement with the crater record fared better when confronted with our proposed dark matter model. A dark matter disk in the usual Milky Way plane with the appropriate density and thickness could adjust the predicted magnitude and time dependence of the galactic plane's tidal force so that both the impact period and the trigger pulse match the data reasonably well.

As a nice bonus, in this way of thinking the look-elsewhere effect of the previous chapter is less compromising than previously thought. We no longer have to think about all possible periods—but only those that take account of the measured ordinary matter density in the galaxy. Armed with the admittedly imprecise measurements of the Solar System and a suitable model of the dark matter disk, we can restrict the range of possible oscillation periods to be only those predictions consistent with existing Milky Way disk density measurements. Matt and I found that with existing data taken into account, the periodic assumption had roughly three times the likelihood of random hits. Though not strong enough statistical evidence to establish the existence of the dark matter disk that we had proposed, the result was promising enough to merit further study.

The best part of this approach is that our knowledge of the galaxy's gravitational potential will continue to improve. Our method, which takes account of all the available information about the galaxy, will become increasingly reliable as more accurate data about the galaxy and the Sun's motion is collected. Scientists today are measuring the distribution of matter in the galaxy. Current satellite observations are recording the positions and velocities of stars, helping us to infer the gravitational potential they experience—namely the potential that binds them to the Milky Way. This in turn will tell us more about the structure of the galactic plane.

In what promises to be some truly exciting results, theory and measurements will tie together Solar System motion with data here on Earth. More data in the future will lead to more reliable predictions, making for yet more trustworthy results.

The next part of the book returns to dark matter models, and closes with the particular model that might explain periodicity in the crater record. The study of periodicity and the Earth's history is an excellent justification to explore—both our immediate visible vicinity and the more ethereal world of dark matter—allowing us to consider the remarkable possibilities for what might be out there invisibly populating our Universe.

PART III

DECIPHERING DARK MATTER'S IDENTITY

16

THE MATTER OF THE INVISIBLE WORLD

The past century's theoretical and observational advances in astronomy, physics, and cosmology have taught us an incredible amount. But the Universe contains a great deal that we have never seen—and likely never will. Various factors account for our limited vision. Many objects are simply too distant to observe. Something very far away won't necessarily emit or scatter enough light to be identified, since any light it did send out would become too widely dispersed and dim.

On top of that, dust or celestial bodies might block our line of sight or obscure our view. Though space probes in distant regions of the cosmos do help surmount these obstacles, no probe has made it to the nearest star—never mind to the nearest galaxy. With their restricted reach and only imperfect resolution, direct probes provide only limited access at best.

Still more factors restrict what we might see. Even if it's in our immediate vicinity, something can be too small to notice. Our visual processing constrains what we can detect without intervening technology. Because we see with visible wavelengths, anything smaller than the wavelength of visible light is beyond our ability to detect with the naked eye. With the latest advances—the Large Hadron Collider in Geneva is the state of the art—we can observe physical processes at smaller sizes than ever before. But even that enormous machine exposes matter down to only about one ten-millionth of a trillionth of a meter. Without further technological progress, sizes and forces relevant to still tinier distances will remain beyond our observational capacity.

But with dark matter, we have an even more airtight excuse not to have seen it. Dark matter doesn't emit or absorb light, which is—let's face it—essential for human vision. Dark matter interacts via gravity, but as far as we can tell in no other discernible fashion. We know of its existence for reasons explained in Chapter 2 and we know some gross features of its properties, but we don't yet know precisely what dark matter actually is. This is what makes it such a compelling subject for research.

In anticipation of our ultimate goal of tying together dark matter and comets, let's now return in this chapter from the study of the Solar System to the subject of dark matter and consider some of the leading possibilities for what dark matter might be.

MODEL BUILDING

Although we are confident it's out there, we don't yet know what dark matter actually is. We know dark matter's average energy density in the cosmos (from the microwave background), its density nearby (from the rotation velocities of stars in the galaxy), that it is "cold"—which is to say, it moves at only a fraction of the speed of light (because we observe structure on small scales in the cosmos), that it interacts at most extremely weakly—both with ordinary matter and with itself (from the lack of discovery in direct searches and from measurements such as that of the shape of the Bullet Cluster), and that it doesn't carry electric charge.

But that's about it. Even if dark matter consists of an elementary particle, we don't know its mass, if it has any nongravitational interactions, or how it was created in the early Universe. We know its average density, but we don't yet know if there is one proton mass per cubic centimeter in our galaxy or 1,000 trillion times the proton mass that is distributed throughout the Universe much more diffusely—

say every kilometer cubed. Small numerous objects and heavier, more dilute ones can both give the same average dark matter density, which is all that astronomers have measured.

Most physicists would bet that dark matter is composed of a new elementary particle that doesn't have the usual Standard Model interactions. Knowing what the particle is means knowing its mass and its interactions and if it is perhaps part of some larger sector of new particles. Many physicists have their favorite candidates, but I won't rule out any suggestion until observations beat me to it.

Fortunately for making progress towards pinpointing the nature of dark matter, one less obdurate reason plays a role in our limited vision. Sheer obliviousness or lack of attention can keep things hidden—even those that we might with current technology have the capacity to observe. Very often we fail to see things that we didn't expect. When I sat in the "lunchroom" on the set of *The Big Bang Theory,* the popular TV show with physicists as lead characters, only a few viewers noticed my presence—I could barely do so myself—even though I appeared very near the lead character, fully within the frame of the TV. (See Figure 35.)

However, obliviousness is something we can correct for. Whereas

[FIGURE 35] As a relatively unnoticed "extra" on the set of *The Big Bang Theory*. (Courtesy of Jim Parsons)

magicians take advantage of this weakness, scientists try to overcome it. Our goal is to identify what we might be missing for lack of attention. Model builders like me try to imagine what might be out there that experimenters haven't yet looked for or realized could be within their grasp. In our models we make guesses for what might underlie and explain the phenomena that we know. With specific models in mind, experimenters can target their searches and data analyses to confirm or rule out any definite suggestions. Even very elusive matter might thereby come into view.

I'm often asked about the criteria I apply when trying to construct a particle physics model. Certainly any good model should be rooted in sound physical ideas that might extend or exploit existing mathematical theories about matter or forces or space. But what are the guiding principles beyond this basic rule?

One that my colleagues and I favor is that models be as economical and predictive as possible. A model with too many moving pieces won't explain anything. A model that is broad enough to accommodate any potential outcome isn't science. Only models that make sufficiently specific predictions to be tested and distinguished from other ideas might ultimately prove interesting.

An additional desirable—though not essential—feature is that its elements connect to existing models. An example would be a dark matter candidate that occurs anyway in models that have been suggested to underlie the Standard Model of ordinary matter. Although such a tie-in isn't guaranteed, such connections are promising in that they avoid additional speculations about an entirely new sector of particles and forces.

Finally, and most essentially, models should be consistent with all known experimental and observational results. Any single contradiction suffices to rule a model out. These criteria apply to all models, including the most popular dark matter models that I'll now discuss.

WIMPS

WIMPs have been the reigning dark matter paradigm among members of the physics and astrophysics communities for several decades. WIMP is the acronym for "weakly interacting massive particle." The word "weak" is not a reference to the weak nuclear force—most WIMP candidates interact even more weakly than the Standard Model's weakly interacting neutrinos. But the interactions are indeed feeble, in the sense that dark matter does not scatter a lot (if at all) as it traverses the Universe.

Moreover, WIMP candidates have mass of about the *weak scale*, which is, roughly speaking, the mass of the recently discovered Higgs particle—accessible at the energy that experiments at the Large Hadron Collider currently explore. To be clear, the Higgs boson is not stable and does have interactions. It is certainly not what dark matter is made of. But other particles with about the same mass might be. If true, not only would dark matter literally be right under our noses, but its identity could prove to be so too—at least for LHC experimentalists.

Support for the WIMP hypothesis is based on a remarkable observation that might be a coincidence or might be a clue to the nature of dark matter. If a stable particle exists with mass comparable to that of the recently discovered Higgs boson, the amount of energy carried by such particles that survive in the Universe today would be about right to account for the energy carried by the Universe's dark matter.

The calculation that demonstrates that particles with this mass are suitable dark matter candidates proceeds from the following observation: As the Universe evolved and its temperature decreased, heavier particles that were abundant in the hot early Universe became much more dispersed. This is because as the temperature dropped, heavy particles annihilated with heavy antiparticles (the equal-mass particles with which particles can annihilate) so that both of them disappeared, but the reverse process where they get created no longer occurred at any significant rate because the energy wasn't sufficient to make them. The consequence was that the number density of heavy particles decreased as the Universe cooled.

If particles had kept to their thermal distribution—the number of them that should be around at a particular temperature—as the temperature dropped, heavy particles would have essentially all annihilated with each other. However, because of the decreasing abundance of heavy particles, the above picture is overly simple. In order to annihilate, particles and antiparticles have to first find each other.* But as their number decreased and they became more diffuse, this meeting became less likely. As a consequence, particles annihilated less efficiently as the Universe grew older and cooler.

The result is that substantially more particles remain today than a naive application of thermodynamics would suggest. At some point both particles and antiparticles became so dilute that they just couldn't find and eliminate each other. How many particles remain depends on the mass and the interactions of the putative dark matter candidate. The intriguing and remarkable conclusion from the appropriate calculations is that stable particles with roughly the Higgs boson's mass happen to be left with about the right abundance to be the dark matter.

We don't yet know if the numbers work out exactly. We will need to know the detailed particle properties to find out. But the fortuitous, albeit rough, agreement between numbers associated with what on the surface appear to be two entirely different phenomena is intriguing and might well be an indication that weak-scale physics accounts for the dark matter in the Universe.

This observation has led many physicists to suspect dark matter is indeed composed of WIMP particles, as these candidates are known. WIMPs have the advantage that because of their connection to Standard Model physics, they are more readily testable than some of the alternative dark matter candidates. WIMP dark matter doesn't interact only via gravity. It also has small nongravitational interac-

---

* Some dark matter particles are their own antiparticle, in which case they can annihilate with other similar particles.

tions with Standard Model particles. Even if those interactions are small, they could still be big enough for very sensitive experiments to record their influence in the so-called direct detection experiments described in the next chapter.

But WIMP searches have so far come up empty-handed. Well, that's not entirely true. Tantalizing hints of detection emerge on a regular basis. But no one is convinced that any of these truly represent dark matter discovery, as opposed to a statistical fluctuation, some problem with the detecting device, or a misunderstanding of astrophysical backgrounds that can mimic the searched-for effect. The evidence is certainly not yet overwhelming.

Yet despite the lack of detection, many scientists just like the idea and still think the coincidence of scales involved in particle physics and dark matter is too good to be accidental. As if this were not overly sanguine enough, many go further and believe in very specific WIMP models, such as those associated with supersymmetry—a theory proposing that for every known particle there is an as-yet unseen supersymmetric partner with the same mass and charges. Given that neither supersymmetric particles nor WIMPs have yet been discovered, even some of the most die-hard supporters have begun to admit some degree of doubt.

As for me, I try to assess the situation as it is at any given point. At a wedding I recently attended, the priest was unusually curious about physics and kept asking me what my gut feeling was for what dark matter would turn out to be. I repeatedly disappointed him by saying I'll let nature decide. As a model builder, I have always had less faith in supersymmetry addressing problems about the Higgs mass, even before the LHC's more recent results, because I knew only too well how challenging it is to make all the pieces fit together. I wouldn't have ruled out supersymmetry and I still don't—that's what experiments are for—but I also wouldn't have said that it was definitely or even likely to be correct.

Similarly, I keep an open mind about dark matter candidates. What I told the priest was true—I don't have a favorite. I try to make testable models because that's the only way we will ultimately learn the answer. As with supersymmetry, the absence of experimental support for WIMPs is making even those who were formerly firmly in the WIMP camp question their confidence that they are on the right track. It is certainly fair in the absence of any experimental support to consider some promising alternatives. I don't know which—if any of them—are realized in the world. But perhaps some other seeming-coincidence of nature constitutes the better clue.

ASYMMETRIC DARK MATTER

One of the most interesting alternatives to WIMP dark matter has various names in the literature, but is most commonly referred to as *asymmetric dark matter.* Models that contain dark matter of this type address another remarkable coincidence that might be accidental or might give us an insight into the nature of dark matter: the amount of dark matter and the amount of ordinary matter are surprisingly comparable.

I recognize that when you first heard that dark matter carries five times the energy of ordinary matter, you might have concluded the opposite—that dark matter overwhelms the energy carried by ordinary matter. However, viewed over the range of possibilities, their energy densities are remarkably similar. The amount of dark matter could have been 700 trillion times more or a googol times less than that of ordinary matter. The evolution of the Universe would of course have been staggeringly different in either of those cases. But nonetheless these ratios were all possibilities.

Yet the energy density of ordinary matter is roughly the same amount as that of dark matter. To say it another way, you can readily see all the slices of the cosmic pie describing the energy carried by dark energy, dark matter, and ordinary matter. There is no slice that you would consider to be a crumb or the entire pie. They are all slices, albeit ones that would lead to different amounts of weight gain if it were indeed a real pie. Without an underlying reason, this is a remarkable coincidence.

In fairness, it is not surprising that the contributions we can observe have comparable sizes. Too small a contribution and a component would not be detectable. The interesting observation is that several components have high enough energy density to contribute comparable amounts today. In principle one contribution might have been a trillion times bigger than the others so that the smaller ones were never observed. But that is not the case. Dark matter and ordinary matter have remarkably similar energy densities.

In asymmetric dark matter models, the similarity in energy densities between ordinary matter and dark matter is not a coincidence. It is a prediction. The coincidence addressed by these models is different than the one that WIMP models are supposed to address, which relates to the amount of energy density left in dark matter after it has partially annihilated. We don't know which of these coincidences—if either—is truly a clue to advancing our understanding. But both types of models are compelling enough to be worth considering, and one of them might even turn out to be right.

In the early 1990s, a number of physicists, including David B. Kaplan, currently director of the Institute for Nuclear Theory at the University of Washington in Seattle, considered this possibility. The idea was revived to account for more recent cosmological measurements in the late 2000s by another David Kaplan (who had formerly been a student at Washington) along with the physicists Markus Luty and Kathryn Zurek. Many other physicists, including me, have worked on this type of model as well.

So what is the idea? To understand the scenario and its motivation, let's take a step back for a moment to reflect on ordinary matter. As alluded to in Chapter 3, the unidentified dark matter is not the only form of matter to present a mystery. Familiar, ordinary matter does too—specifically, the amount of it we find in the Universe today. The energy of most ordinary matter lies in protons and neutrons, which are a form of *baryon*—matter that is ultimately composed of elementary particles called quarks. Had ordinary matter, consisting mostly of baryons, been distributed according to the simplest early Universe scenario—annihilating away as the Universe cooled—it would have far lower density than that which we currently observe.

A critical feature for our Universe—and ourselves—is that in contrast to standard thermal expectations, ordinary matter sticks around and survives in sufficient quantities to create animals and cities and stars. This is possible only because matter dominates over antimatter—there is a matter-antimatter asymmetry. If the amounts had always been equal, matter and antimatter would have found each other, annihilated, and disappeared.

Clearly at some point in the Universe's evolution, the amount of matter surpassed the amount of antimatter. Without such an excess, too much of the matter would be gone by now. But we don't yet know how this occurred. Matter-antimatter asymmetry happened only with some special interactions and conditions in the early Universe. Some process must have gone out of thermal equilibrium—proceeding too slowly to keep up with the Universe's expansion—or else matter and antimatter particles would be created in equal numbers. Moreover, symmetries that might have seemed natural can't apply when the excess matter gets created.

We don't know what gave rise to either the symmetry breaking or the departure from thermal equilibrium, though there are suggestions for both in the context of Grand Unified Theories, models of leptons (particles like electrons and neutrinos that don't experience the strong interactions), and supersymmetry. No one will know which—if any—of these models is right until someone finds observational evidence. Unfortunately, many of these scenarios don't have readily observable consequences.

Even so, we can be confident that at some point a process called *baryogenesis* occurred, in which an excess of matter over antimatter—a matter-antimatter asymmetry—gets created. Without baryogenesis, none of us would be here to tell even this partial tale.

Asymmetric dark matter models suggest that since the dark matter energy density is so similar to that of ordinary matter, maybe dark matter too got created in a related process involving a dark matter–dark antimatter asymmetry. Matthew Buckley—who was a Caltech postdoctoral fellow when we worked on this topic—and I labeled this process *Xogenesis,* playing with the idea of dark matter as the unknown quantity, X. The truly compelling aspect of any of these models is that not only do they allow for dark matter creation in analogy to ordinary matter—the two are actually related in the most interesting examples. If dark matter and ordinary matter have some form of interaction, even a rather feeble one and perhaps one that was stronger earlier on, the energy density of ordinary matter and dark matter should be comparable, addressing the coincidence we want to explain. This is the strongest motivation for believing these models might turn out to be correct.

AXIONS

WIMP and asymmetric dark matter models are both general paradigms. WIMP-type models involve a weak-scale stable particle and asymmetric dark matter models suggest an asymmetry of dark matter over anti–dark matter. Both ideas encompass a wide variety of implementations that might include distinguishing particles and interactions.

Axion models deal with a more restrictive scenario. An axion appears only in models associated with a very specific issue in particle physics known as the strong *CP* problem, where *C* stands for charge and *P* stands for parity. *C*- or charge-conservation says that interactions of positively and negatively charged particles are closely related. *P*- or parity-symmetry says that no physical laws should distinguish left from right, so that, for example, particles spinning to the left and spinning to the right should have identical interactions. Yet not only do nature's interactions independently violate *C* and *P*. They also violate the combination together. That is to say, the violations in *C* and *P* don't compensate each other.

However, for yet unknown reasons, *CP* violation—as the combination of *C* and *P* symmetries is known—happens only in some processes. The absence of an explanation for why *CP* should restrict interactions in some cases but not in others within the context of

Standard Model physics is known as the *strong CP problem*. Axions feature in a proposed solution to this conundrum.

I mention all this for completeness. Without some background in particle physics or an entire book on the subject, I'm aware that these ideas can be difficult to grasp. Fortunately, to understand the axion's cosmological implications and its role as a potential dark matter candidate, you don't need to follow the particle physics specifics. The cosmological predictions depend only on an axion being extremely light and having extremely weak interactions.

You might think such features would render axions harmless—in fact that is what most physicists did think initially. But in a remarkable paper, the particle theorists John Preskill, Frank Wilczek, and Mark Wise explained why this is not necessarily the case. These authors showed that axions are so weakly interacting and so light that the number of axions wouldn't affect the energy of the early Universe. No physical process determined how many of them should be around. Their presence would not have affected the Universe's evolution until it cooled sufficiently.

Because of the axion density's irrelevance early on, at the time when axions do finally make a difference, their number is very unlikely to be the one the Universe would favor—the number that minimizes energy, for example. The Universe would therefore then find itself with a large number of axions in an enormous condensate—so many that even though the axion is very light, the axion condensate's energy is large. In a surprising turn, axions cannot interact too weakly or the Universe would have more energy than is allowed.

The above considerations restrict the allowed range of axion interactions. But turning this observation on its side: if the interactions are weak but not too weak, axions might carry a large energy density—just not necessarily so large as to contradict observations. In fact, if the interaction strength is just right, dark matter could be composed of axions that carry precisely the measured dark matter energy density.

Axions have a mass that is very different from either of the dark matter candidates described previously. Whereas both the other suggestions involved dark matter particles with mass near the weak scale or maybe a hundred times less, axion scenarios involve extremely light particles, with mass at least a billion times smaller.

Furthermore, axions interact very differently from the other dark matter candidates. The combination of cosmological and astrophysical constraints place axion models in a very narrow window of permissible masses and interaction strengths. The interactions can't be too weak or there would be too much axion energy density. But they also can't be too strong or we would have seen axions through direct production in particle experiments or in stars. That's because axions that interact sufficiently strongly would be produced in stars and could thereby cool them down. The observed rate of supernova cooling shows no sign of nonstandard contributions, and thus limits the strength with which axions can interact.

Theoretically, in light of the constraints, I find axion models a little odd in that the window of interactions that is experimentally permitted is pretty random with no clear relation to any other physical processes. I'm a little doubtful that experiments searching for axions will find a positive result, but many of my colleagues are more optimistic. Ongoing searches for axions involve their admittedly very feeble interaction with light. Axion detectors placed in huge magnetic fields look for measurable radiation that the axion's interaction with a magnetic field would produce. Only time—and such experiments—will tell whether or not axions exist in nature, and, if they do, whether they indeed constitute the dark matter.

NEUTRINOS

A common aspect of all the models I've presented is that they include some connection between ordinary matter and dark matter—hinted at either by the coincidence of mass scales for WIMPs, by the proximity of energy densities for asymmetric dark matter models, or by a proposed solution to the strong CP problem in the case of axions. Axion models were hypothesized for a particle physics reason, but might nonetheless account for dark matter. WIMPs too feature as part of proposed particle physics scenarios, such as supersymmetry. Asymmetric dark matter models might also exist in the framework of existing theories, but, despite the interactions between dark matter and ordinary matter, usually an asymmetric dark matter candidate is an extraneous addition to the theory.

However, it might well be that dark matter—or at least some of it—interacts purely gravitationally. Dark matter might also have some of its own forces and interactions, not experienced by our matter.

But before proposing the existence of an independent dark matter sector, physicists first considered whether any type of ordinary matter might itself have the properties that could allow it to be the dark matter. The question was whether anything in the Standard Model or something that was composed of Standard Model particles could be a suitable dark matter candidate, even without the inclusion of additional particles.

One of the earliest such suggestions concerned a type of elementary particle called a *neutrino.* In the radioactive process known as *beta decay,* neutrons decay to protons and electrons and neutrinos (technically, to their antiparticles, known as *antineutrinos*). Neutrinos are particles that—like electrons, and their heavier counterparts known as *muons* and *taus*—don't experience the strong nuclear force. Also, they carry zero electric charge so they don't directly experience electromagnetism either. One of the most interesting properties of neutrinos is that (aside from gravity, which all particles experience—albeit at an extremely tiny level) they interact directly only through the weak force. The other is that they are known to be very light, at least a million times lighter than an electron.

Because they interact so feebly, neutrinos originally seemed like promising dark matter candidates. But by now that hope has been quashed for several reasons. The neutrinos of the Standard Model interact via the weak nuclear force. But for anything sufficiently light that has weak force interactions, the direct detection experiments discussed in the next chapter—which have so far come up empty-handed—would have already seen something. On top of that, ordinary neutrinos that we know to exist can't be the dark matter since their energy density is far too small. For dark matter's energy to be matched by that of neutrinos, neutrinos would have to be much heavier.

In fact, light neutrinos would be a form of *hot dark matter,* which travels at nearly the speed of light. Hot dark matter would wash out structure on scales smaller than a supercluster. Yet we observe galaxies and galaxy clusters. So this would be a problem.

So neutrinos don't work in the context of the Standard Model of particle physics. Physicists subsequently considered modifications to the Standard Model that also didn't work. Particles that interact like neutrinos—even modified versions of neutrinos—don't give rise to the nice story of structure formation that was outlined in Chapter 5.

In principle, hot dark matter could still be okay if the smaller structures that we know to exist weren't formed directly but had formed from the fragmentation of bigger ones. But this scenario can be worked out numerically and the predictions don't match observations. So despite the occasional hints of new light neutrinos, and the appearance of headlines about dark matter neutrinos, neutrinos really aren't the dark matter. Neutrinos can at best account for a small fraction of the existing dark matter density. This is why physicists concentrate their attention on *cold dark matter* scenarios, which suggest more slowly-moving dark matter candidates—which are usually heavier. *Hot dark matter* scenarios—what you get when you have light and very fast-moving particles such as neutrinos—are ruled out.

MACHOS

Finally, let's consider the superficially more likely possibility that dark matter isn't composed of novel elementary particles and is instead composed of nonburning (and hence non-light-emitting), nonreflective macroscopic structures made of ordinary stuff. We wouldn't see these objects for the same reason we don't see in a dark room. It's not that the stuff around you doesn't interact with light at some level. It's just that there isn't enough light around for you to see. Before accepting the existence of dark matter, most anyone—scientifically inclined or not—would want to know why this seemingly obvious possibility isn't true.

Dark objects like these are known collectively as MACHOs, an acronym for massive compact halo objects—their role as an alternative to WIMPs being the not-so-subtle reason behind the name. Since MACHOs emit little or no detectable light, they could appear to be hidden and dark even though they are composed of ordinary matter. MACHO candidates include black holes, neutron stars, and brown dwarfs.

As introduced earlier, black holes are very tightly gravitationally bound states of matter that don't emit or reflect light. *Neutron stars*—possibly produced in supernova collapse—are the remnants of massive stars that don't have quite enough matter to become black holes but instead condense into a state with an extremely dense neutron core. *Brown dwarfs* are objects that are bigger than Jupiter but smaller than stars, so they couldn't ignite via nuclear fusion, but instead heat up only through their gravitational contraction.

The above astrophysical objects certainly seem like reasonable possibilities. But even before more recent observations severely restricted the possibility, MACHOs were considered unlikely. You might recall from Chapter 4 that one of the early tests of the standard Big Bang scenario was the creation of nuclei in the early Universe—the process known as primordial nucleosynthesis. But that scenario works only for some definite range of values for the energy density in ordinary matter. Most MACHO models contain too much ordinary matter to yield the correct nuclei abundance predictions. On top of that, even if ordinary matter formed these compact objects, understanding how they ended up in the halo rather than in the galactic disk poses another important challenge.

Even so, astrophysicists kept an open mind. Dark matter is an extraordinary proposal so it's worth ensuring that every conventional explanation is ruled out. In the 1990s, physicists searched for MACHOs through a process known as *microlensing*. According to this beautiful and subtle idea, MACHOs moving through space occasionally pass in front of a star. Because light rays bend around a MACHO (or any other massive object), it acts as a lens whose gravitational influence temporarily amplifies the light from the star, making it appear brighter before it returns to normal. Of course this has to happen on a time scale that is short enough to be observed and the change in magnitude has to be a big enough effect to be detected. But astronomers using this method could show that MACHOs in the range between about a third the mass of the Moon to about 100 times the mass of the Sun cannot be the dark matter, thereby ruling out many MACHO candidates.

Though MACHO searches ruled out neutron stars and white dwarfs, black holes in a narrow mass window remained a possibility. Aside from the lack of theoretical reasons to believe there would be the right amount of black holes in any given mass range, the gravitational disruptions caused by black holes as well as the duration of a black hole's existence do nonetheless place further limits. This is because black holes that are too small would have already decayed away by emitting photons in a process known as Hawking radiation (named after the physicist Stephen Hawking, who first proposed it), whereas those that are too big would have observable effects that haven't been seen. These include disruptions of binary systems, scattering that can heat up and widen the plane of the Milky Way, black holes accreting other matter and radiating, and the effect of gravity waves associated with black holes on precisely measured pulsar timing. Putting the constraints together might permit black hole masses from about a millionth of a lunar mass to a lunar mass, but black holes outside this mass range are ruled out. Detailed measurements of neutron star properties might rule out even this remaining, limited range.

Even if a very small window for black hole masses does persist, it would be hard to fathom why only black holes in this one particular mass range would be created and would have survived in sufficient quantity. It's fair to consider the possibility. But based on nucleosynthesis and model-building constraints, black holes—especially ones that were created only by ordinary matter—are extremely unlikely to be the dark matter.

WHAT TO DO?

The above models contain the most commonly discussed candidates for dark matter—the ones that most physicists think of as reasonable possibilities. But they are almost certainly not the only options. Though some of these ideas are still promising, we have good reason to be skeptical of any specific model or property until it is experimentally confirmed.

On the other hand, we can be very confident that dark matter exists—even if we don't yet know what it is. This is presently a good time for both theorists and experimenters to reassess and to try to consider a more complete range of options. Most of them will involve different types of search strategies. Alternative models will be useful in planning these out.

But before getting to some newer ideas, I'll first review some of the existing dark matter search techniques in order to more knowledgeably assess the current situation. We'll see that the abundant data in astrophysics coupled with the lack of detection of previously proposed models gives experimenters and observers good reason to look beyond these older and more developed dark matter searches.

17

HOW TO SEE IN THE DARK

The Brown University professor Richard Gaitskell, a principal investigator and co-spokesperson of LUX—a major dark matter detection experiment—spoke at Harvard in December 2013. In the colloquium he gave to the many rapt physics department members in attendance, he gleefully described how he and his collaborators hadn't yet discovered dark matter. The measure of his experiment's success was, curiously enough, that it had ruled out many of the dark matter candidates that a large class of models and even some (now) spurious experimental results had suggested. Yet despite the disappointing physics news that dark matter had not yet been found—by his or any other experiment—Gaitskell was justifiably elated. The very challenging experiment he and others built had worked as well as he had hoped. It wasn't his fault that nature hadn't cooperated and provided a dark matter candidate with a mass and interaction strength that his experiment could find.

This was only the first set of results from the LUX experiment, which continues to run and collect more data—yet it had overtaken the results from the older, more well-established experiments right out of the gate. Gaitskell and his collaborators had created such a clean environment that the experiment's very first results were sufficiently trustworthy to supersede older findings. In a setting where radioactivity from a rogue experimentalist's misplaced fingerprint can contribute billions of times more "signal" than the sought-after feeble imprint of a dark matter particle, Gaitskell's experiment had done spectacularly well. The clean and reliable data the experiment had collected definitively established that his apparatus did precisely what it had been designed to do—extremely sensitive searches and trustworthy rejections of any misleading signals.

Lots of data gets collected through the latest technology today, and not all of it is about people's consumer preferences. Data currently being amassed will lead to advances in particle physics, astronomy, and cosmology—as well as other fields of science. Although no experiment has decisively discovered dark matter, many experiments have presented tantalizing results. Sometimes an experiment like Gaitskell's comes along and rules out many of the promising possibilities that previous experiments with less decisive measurements had suggested. His and other experiments continue their searches in the hopes that they will soon find a more robust signal that will represent a true discovery.

However, the search for dark matter is a daunting task. Because gravity is such a weak force, the searches for the particles that compose dark matter need to invoke precisely those interactions that we don't yet know that dark matter experiences. If dark matter interacts solely gravitationally, or via new forces not experienced by ordinary matter, conventional dark matter searches will never find it. Even if Standard Model forces do act on dark matter, we still can't be sure that the interactions are strong enough for current experiments to detect.

Today's searches rely on a leap of faith that dark matter, despite its near invisibility, has interactions that are sufficiently substantial for detectors built from ordinary matter to register. This is based partially on wishful thinking. But optimism is also rooted in the implications of the WIMP models discussed in the previous chapter. Most WIMP dark matter candidates should interact with Standard Model particles a tiny bit—small indeed, but at rates that could potentially be observable with the very precise experiments now in operation. Searches have reached the point where most WIMP models will be confirmed or ruled out once these experiments present their final results.

When investigating alternative dark matter models in the chapters that follow, I'll present some of their very different observational implications. But this chapter's focus is WIMP dark matter, and the three favored approaches for seeking it. (See Figure 36.) Dark matter is elusive, but experimenters have been intrepid in their search for its subtle observable consequences.

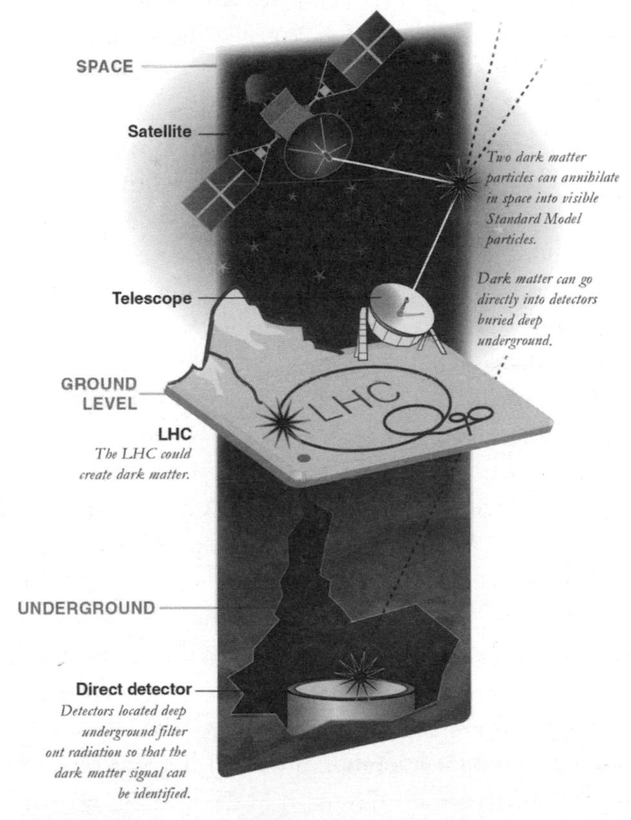

[FIGURE 36] Searches for WIMPs take a three-pronged approach. Underground detectors look for dark matter that directly hits target nuclei. Experiments at the LHC might find evidence of dark matter that the LHC created. Satellites or telescopes look for evidence of dark matter that annihilates into visible matter in indirect detection searches.

DIRECT DETECTION EXPERIMENTS

The first class of experiments that look for WIMPs fall into the category of *direct detection.* Direct detection experiments involve enormous, extremely sensitive apparatuses on Earth, whose large detection volume is designed to compensate for dark matter's (at best) minuscule interaction strength. The idea behind these searches is that dark matter should travel through the material of a detector until it hits a nucleus. The interaction would then provide a small amount of recoil heat or energy that can in principle be measured either with a very cold detector or with a very sensitive material that is designed to absorb and record the tiny heat that might be deposited. If a dark matter particle passes through a direct detection apparatus and hits and subtly ricochets off a nucleus, the experiment might record the tiny change in energy that would be the sole potentially measurable evidence of its passage. Although the chance of any individual interaction is very small, the probability of success improves with larger size and better sensitivity, which is why the experiments are so big.

Cryogenic detectors are very cold devices with crystal absorbers such as germanium. They respond to the small amount of heat by using *SQUIDs*—superconducting quantum interference devices—which are built into the detector. These devices lose superconductivity and register a potential dark matter event if even a small amount of energy hits the very cold superconducting material that is built into it. Experiments in this category include the Cryogenic Dark Matter Search (CDMS), the Cryogenic Rare Event Search with Superconducting Thermometers (CRESST), and the Expérience pour Détecter Les Wimps en Site Souterrain, which is French for "Experiment to Detect WIMPS in an Underground Site" (EDELWEISS). The names are a handful but most physicists use only the more tractable acronyms.

Cryogenic detectors are not the only type of detector used for direct detection. The other type—which is rapidly increasing in importance—employs noble liquids. Even though dark matter doesn't directly interact with light, the energy added to an atom of xenon or argon when hit by a dark matter particle can potentially generate a flash of characteristic scintillation. Experiments of this sort include the xenon-based experiments XENON100 and LUX (the Large Underground Xenon Detector)—the experiment I mentioned above—as well as the argon-based detectors named ZEPLIN, DEAP, WARP, DArkSide, and ArDM.

Both XENON and LUX will have bigger and better upgrades in the coming years—XENON1T and the LUX-ZEPLIN collaboration. To give an idea of the progress, the "100" in the original XENON name was a rough measure of kilograms whereas 1T stand for "one ton." LUX-ZEPLIN will be even bigger, with a five-ton *fiducial* volume, which is the region used for dark matter detection.

Both cryogenic and noble gas detectors are designed to record the tiny energy a dark matter particle might deposit. However, impressive as this may be, detecting a small change in energy won't suffice to establish that a dark matter particle has passed through. Experimenters also need to establish that they have recorded the desired signal and not just background radiation, which also can deposit small amounts of energy that might mimic dark matter and that interacts much more strongly with ordinary matter than dark matter ever could.

This is tough. Radiation, from the perspective of a sensitive dark matter detection apparatus, is everywhere. Cosmic ray muons—heavier partners of electrons—can hit rock and create a splash of particles, as can some neutrons that mimic dark matter. Even making reasonably optimistic assumptions about the mass and interaction strength of the dark matter particles, background electromagnetic events dominate over the signal by at least a factor of a thousand. And this estimate doesn't take into account all the primordial and man-made radioactive substances that are present in the air, the environment, and the detector.

The scientists who design these apparatuses know this all too well. The name of the game for astrophysicists and dark matter experimenters is *shielding* and *discrimination*. In order to protect their detector from dangerous radiation and distinguish potential dark matter events from uninteresting radiation scattering in the detectors, experimenters look for dark matter deep underground in mines or below mountains. Cosmic rays should then hit the rock surrounding a sufficiently buried detector instead of the detector itself. Most of the radiation will be screened, whereas dark matter, which has much weaker interaction strength, will make it to the detector unimpeded.

Fortunately, plenty of mines and tunnels that have been built for commercial purposes are available to house such experiments. Mines exist in part because—as noted earlier—heavy elements sink to the center of the Earth, but some fraction occasionally rises up in ore deposits underground. The DAMA experiment, along with experiments called XENON10 and the bigger XENON100, as well as CRESST, a detector that uses tungsten, takes place in the Gran Sasso National Laboratory, situated in a tunnel in Italy about 1,400 meters underground.

A 1,500-meter-deep cavern in the Homestake mine in South Dakota, originally built for gold excavation, is home to the LUX experiment. The Homestake mine is famous in physics circles as the location where another impressive detection—neutrinos from the Sun—took place and provided the first real clue that neutrinos had nonzero masses. The CDMS experiment is in the Soudan mine that is located about 750 meters underground. The Sudbury mine in Ontario, Canada—the mine created to dig up the metals concentrated in that region by an enormous asteroid that hit about two billion years ago—is home to several dark matter experiments too.

Still, all that rock above the mines and tunnels is not enough to guarantee that the detectors are radiation-free. The experiments further protect the detectors in a variety of ways. The shield I find most amusing is old lead taken from a sunken French galleon. Lead is a dense absorbing material, and old lead has already rid itself of whatever radiation it initially contained, so it is effective at absorbing radiation without providing new sources of its own.

Other, more technologically advanced shielding comes, for example, from polyethylene that lights up if something interacts too strongly to be dark matter. In the noble liquid detectors, such as those using xenon, the detector itself functions as a shield. The absorbing region of these xenon detectors is so large that the experimenters omit the outer region, which is used only to block radioactive backgrounds when they record the potential signal events that come solely from the inner region.

Discrimination is important too. Particle physicists use a different term, *particle identification (particle ID)* for this requirement. Maybe the particle physics term is more PC, though "ID" requirements are politically loaded these days as well. Whatever the name, discrimination—as opposed to shielding—is about distinguishing the electromagnetic radiation that nonetheless makes it through from potential dark matter candidates. By measuring both ionization and the initial scintillation, experimenters can distinguish signal from background radiation.

The DAMA experiment, a scintillation experiment located in the Gran Sasso National Laboratory in Italy, has been reporting a signal for some time now. However, because the experiment lacks discrimination between signal and background—relying solely on timing information—and because no other experiment has reproduced the result, most physicists remain skeptical about the signal's authenticity.

Other experiments also have recorded potential signals, but with very few events—and those only at low energies. There is good reason to be suspicious of these results too. You will recall that the detectors measure recoil energy. When that energy is too small, the detector can't record it since it's below the sensitivity of the apparatus. The lowest-energy events are closest to the difficult-to-access low-energy boundary. So skepticism about any potential low-energy signal is warranted until more data arrive or some other experiment confirms the potential observations.

INDIRECT DETECTION

Direct detection experiments that look for dark matter that passes through Earth might succeed in their mission and discover dark matter particles. But another promising search strategy is to look for the signal that would arise if dark matter particles annihilate with dark matter antiparticles (or the same type of particle when it can annihilate with itself), transforming the dark matter particle's energy into other—hopefully visible—types of matter. Dark matter annihilation probably doesn't happen very often, since dark matter is so dilute. But this doesn't mean that it doesn't happen at all. It depends on what the nature of dark matter turns out to be.

If and when annihilations occur, experiments on Earth or in space might find the particles that emerged from the annihilation in what is known as *indirect detection*. Such searches look for the particles that were created after the annihilating dark matter particles disappeared. If we are lucky, those emerging particles will include Standard Model particles and antiparticles, such as electrons and their antiparticles—positrons—or pairs of photons, all of which detectors on Earth and in space could potentially observe. Antiparticle and photon signals are the most promising search targets associated with the indirect detection of dark matter because of the rarity of antiparticles in the cosmos. Photons can be useful too, since photons originating from the annihilation of dark matter will have a different energy and spatial distribution from photons created by astrophysical backgrounds.

Most of the instruments that search for these Standard Model products of dark matter annihilation weren't originally designed as dark matter detectors. The primary goal of the majority of telescopes and detectors that are out in space or on the ground is to record light and particles from astronomical sources in the sky. The objective is a better understanding of the stars, pulsars, and other objects, which, to a dark matter experimentalist, are the astrophysical background that might mimic a dark matter signal.

Viewed differently, this similarity of particles emitted by astrophysical background sources and putative dark matter annihilations tells us that the observations of existing telescopes can potentially tell us about dark matter too. If astrophysicists understand the more conventional sources of such particles, they can distinguish them from an excess due to dark matter. Despite potential ambiguities in interpretation, indirect dark matter searches could succeed if conventional sources are sufficiently well understood to guarantee that they wouldn't suffice to account for what might be found.

One such indirect detection experiment exists on the International Space Station. The Nobel Prize winner Sam Ting from MIT had the clever idea of putting a particle detector there to look for positrons and antiprotons. The Alpha Magnetic Spectrometer (AMS) is essentially a particle detector in space. It has extended the search that the Italian-led PAMELA satellite (the pretty name is also an acronym), which first reported results in 2013, had already done.

Although the data had looked intriguing at first, dark matter as an explanation is now strongly disfavored because, among other things, the PAMELA and AMS signals would require so much dark matter in the early Universe that its distorting effects on the microwave background should have been seen by the Planck satellite. The initially surprising outcome now seems to be merely an indication that astrophysicists have a lot to learn about astrophysical sources such as pulsars. As long as conventional sources stand a chance of explaining the signal, no convincing argument for dark matter can be made.

Dark matter might also annihilate into quarks and antiquarks or into gluons—particles that interact via the strong nuclear force. In fact, most WIMP-type models predict this to be the most probable Standard Model outcome. The astrophysical backgrounds for the most obvious search target—antiprotons—is big, but those for low-energy antideuterons, which are very weakly bound states of antiprotons and antineutrons, are much lower. Experiments might have a chance of discovering dark matter through their annihilations into these low-energy states. The balloon-based experiment GAPS, scheduled to launch from Antarctica by 2019, will look for this signal.

The uncharged particles called neutrinos that interact only via the weak force could also help with indirect dark matter detection. Dark matter might get trapped in the center of the Sun or the Earth, increasing the dark matter density over its usual value and enhancing the probability of annihilation. The only particles that could escape and potentially be detected would then be neutrinos, since—unlike other particles—they are too weakly interacting to be blocked from escaping by their interactions. Detectors on the ground named AMANDA, IceCube, and ANTARES are looking for these high-energy neutrinos.

Other ground-based detectors look for high-energy photons, electrons, and positrons. HESS, the High Energy Stereoscopic System, located in Namibia, and VERITAS, the Very Energetic Radiation Imaging Telescope Array System, in Arizona, are large arrays of telescopes on Earth that look for high-energy photons from the center of the galaxy. The next generation very-high-energy gamma-ray observatory, the Cherenkov Telescope Array, promises to be even more sensitive.

But probably the most important indirect detection searches in the last couple of years have been performed by the telescopes on the Fermi gamma-ray space observatory—informally known as Fermi—named after the great Italian physicist for whom "fermions" are also named. The Fermi observatory sits on a satellite that was launched early in 2008, orbiting the sky every 95 minutes at a height of 550 kilometers above the Earth. Photon detectors that sit on Earth have the advantage that they can be much bigger than a satellite in the sky. But the very precise instruments on the Fermi satellite have better energy resolution and directional information, are sensitive to photons with lower energies, and have a much wider field of view.

The Fermi satellite has been the source of a lot of interesting speculation of late about dark matter. Several hints of signals have emerged since it began operation, none of which are decisive but all of which have led to interesting insights into what dark matter might be. The strongest signal at this time is one that the physicist Dan Hooper from Fermilab (the Fermi National Accelerator Laboratory, located in Batavia, Illinois, near Chicago) has been advocating. He and his collaborators observed that a careful study of the diffuse photon emission near the galactic center shows an excess over what is expected from astrophysical backgrounds.

As with the earlier, surprising positron result, the data fairly decisively show an excess over expectations. The question again is whether the missing ingredient is some neglected astrophysical source or something truly exciting like dark matter. Astronomers are still at work, trying to determine the answer. As of now, neither explanation seems entirely straightforward or convincing.

Another signal that has emerged in photons and has no known explanation from conventional astrophysical sources comes in the form of an X-ray line at a few keV,* about a hundredth of the energy carried by an electron. The peculiar feature of the observation is that it is a line, which means the excess photons occur at a definitive energy with very little energy spread. Mind you, atomic and molecular transitions between different energy levels give rise to similar lines and the signal is not exceptionally strong, so it's far from clear this is a real discovery. This lack of convincing evidence hasn't stopped some research based on axions or decaying dark matter as possible sources. We won't know until further data or theoretical work shows whether it was a fluctuation or background or a true discovery of something new.

The final putative signal that I want to mention, because it instigated the research I will shortly discuss, is a photon signal with 130 GeV of energy, which early Fermi satellite data had initially seemed to suggest. The signal was certainly intriguing, given the similarity of the observed energy to the mass of the Higgs boson, which is about 125 GeV. In light of the lack of a good astrophysical explanation, some astronomers suggested the signal might have originated from annihilating dark matter.

I will say right up front that the evidence didn't stand the test of time—or further data—and is now discredited. But in trying to

* electronvolts (eV), the unit of energy most used by particle physicists. keV, kiloelectron volts represents one thousand eV, whereas GeV, gigaelectron volts—a unit frequently used in discussions of physics at today's high-energy accelerators—is one billion electron volts.

explain how the signal could possibly come about, my collaborators—Matt Reece, JiJi Fan, Andrey Katz—and I ended up exploring an intriguing class of models that we almost certainly never would have stumbled across otherwise. As with many interesting scientific developments, the model turned out to be interesting for reasons beyond the initial motivation and led to the dark disk model I will soon explain.

DARK MATTER AT THE LHC

Although looking like a less promising possibility these days, WIMPs might also show up at the Large Hadron Collider, the giant accelerator near Geneva that circles under the French-Swiss border. Protons circulate in opposite directions around a 27-kilometer circumference ring to collide with each other at high energies. The LHC spans a range of energies that allowed for the Higgs boson's creation and discovery, and might also allow for the production of other, hypothetical particles such as a stable, feebly interacting WIMP. If so, its interactions with Standard Model particles might permit its detection at the LHC.

Even if new particles are found at the LHC, supplementary evidence—such as that from dedicated dark matter detectors on the ground or in space—will be required to truly establish that a newly discovered particle constitutes dark matter. Nonetheless, finding WIMPs at the LHC would certainly be a major accomplishment. We might even learn detailed properties of the dark matter particle that would be very difficult to study with either of the other detection methods.

However, dark matter particles wouldn't interact a lot with the protons that collide at the LHC since dark matter has such tiny interactions with ordinary matter. Even so, other particles might be produced that decay into them. The question then would be how to establish that this has occurred, since dark matter doesn't interact with the detector and therefore in itself leaves no visible evidence.

One place to look would be in the decay of charged particles. The charged particles won't decay solely into neutral dark matter particles because this process wouldn't conserve charge. By detecting the additional charged particles that must be present in the final state, which won't carry the same energy and momentum of the initially decaying particle since unseen dark matter has carried energy and momentum away, the existence of a feebly interacting particle with a particular set of interactions might be established.

The sign that dark matter had been produced would be precisely the energy that experiments failed to detect, along with an agreement of predictions for event rates and signals with data. Unless the laws of physics are radically different than anyone believes, the apparent lack of energy and momentum conservation could only be interpreted as the production of an undetected particle, which might then be attributable to dark matter.

WIMPs, despite their tiny interaction with ordinary matter, can also be directly produced in pairs. Two colliding protons might sometimes produce two WIMPs, in the reverse process of two WIMPs annihilating to produce ordinary matter—the calculation that leads to the relic abundance result. How often this occurs depends on the particular model—after all, WIMPs didn't necessarily annihilate into protons, so the reverse process isn't guaranteed either. But for many models, this can be a good way to search.

Again, experimenters have to address the problem that dark matter itself is never detected—only other particles produced along with it. But experimenters can see events where a single particle like a photon or a gluon (the particle that communicates the strong nuclear force among quarks) gets produced along with the dark matter, and theorists have demonstrated that these searches could in principle give rise to a big enough signal.

So far, LHC studies have not found anything indicative of dark matter's production. Physicists don't know whether this is because the machine's energy is a little too low or because the theoretical considerations suggesting that additional particles will be found at these energies are misguided. But there is still a reasonable chance of finding additional particles at the energies the LHC collisions will produce. Maybe one of these will be dark matter.

LOOKING FOR LESS WIMPY DARK MATTER

WIMPs, unlike Obi-Wan Kenobi, are not our only hope, though as far as these detection methods are concerned, they are in many ways our best one. Direct detection works only when there is some interaction between Standard Model and dark matter particles, and WIMP models guarantee that possibility. Furthermore, thermal production guarantees equal amounts of dark matter and anti–dark matter (or that dark matter is its own antiparticle) so that annihilation is not out of the question either. But what about the other suggestions for dark matter? How do we look for those?

Unfortunately, any other dark matter candidate that hasn't already been ruled out will most likely be even more difficult to find. The search strategy has to be specific to the particular model. Accessibility with current technology is not necessarily guaranteed. Maybe we'll be lucky and dark matter won't be transparent—it will be diaphanous, just barely visible to these methods that optimistically assume some interaction with Standard Model forces. But, given the uncertainties, my opinion is that it's time to focus more on detection through the one force we know is present for dark matter: gravity. Dark matter interacting with itself or with other invisible matter might not show up directly, but we will now see that its interactions might reveal themselves through the exact distribution of mass in the Universe.

18

SOCIALLY CONNECTED DARK MATTER

Urbanization has been vital to many of the advances in modern life. Put enough people together and ideas bloom, economies flourish, and abundant benefits emerge. Cities develop organically as they expand—becoming yet more attractive as people move in, create jobs, and establish better working and living conditions. But once a city becomes too dense, expensive housing, crime, or other urban predicaments frequently drive people out into more sparsely settled neighborhoods, or even farther away—outside the city altogether. The rest of the city might grow as planned, but overly optimistic real estate developers' ambitions will be frustrated by the inadequately populated high-rise buildings of the inner cities that defied earlier projections of very rapid growth. And without stable urban centers, suburban communities won't flourish either, in which case mall developers will be disappointed too.

It turns out the same general pattern might apply to the growth of structure in the Universe. I've explained our current understanding of dark matter and the many observations and predictions that convince us that dark matter and ordinary matter interact with each other at most very weakly. Numerical simulations based on dark matter that interacts solely gravitationally predict the size, density, concentration, and shapes of galaxies and clusters of galaxies. As with the predictions for large-scale urban growth, predictions of large-scale structure in the Universe agree with observations extremely well.

But precise numerical simulations assuming the usual dark matter properties don't always match observed density profiles on smaller scales. The central regions of galaxies and galaxy clusters and the number of smaller dwarf galaxies near the Milky Way don't agree with theoretical predictions. As with less densely populated inner cities and under-developed suburbs, the projected densities for the centers of galaxies and the predicted numbers of satellite galaxies are both too high. The dwarf galaxies in Andromeda and other galaxies don't match the predicted spatial distribution either.

It might turn out that simulations are inadequate or observations are still too incomplete. But disagreements between predictions and observations of structure on small scales might also suggest that dark matter is different than is currently assumed. Perhaps dark matter is not so very weakly interacting after all.

Though interactions between dark matter and ordinary matter are known to be tiny, the interactions between a dark matter particle and another dark matter particle can be fairly large. Such self-interactions are less constrained by the absence of dark matter detection, which tests only for interactions between dark matter and ordinary matter. They might be big enough to merit some attention.

Though we can now confirm the basic ideas about the growth of structure in the Universe, possible discrepancies indicate that the science hasn't advanced sufficiently to consider the topic closed. For researchers, this is the optimal situation. We are bound to learn a lot—whatever the resolution. This chapter explains the issues with the Universe's structure on smaller scales, and why self-interacting dark matter could address them.

SMALL-SCALE ISSUES

Chapter 5 explained how gravity acting on dark matter lays out a blueprint that determined the Universe's structure. Dark matter in the early Universe developed increasingly large density fluctuations and galaxies grew in the denser regions, which exerted the most gravitational pull. Once formed, galaxies merged into clusters located along sheets and filaments and these formed the scaffolding on which other structures were built. Although the details of each individual galaxy or cluster depend on the unknown initial state, astronomers predict the overall statistical properties of galaxy and galaxy cluster distributions, and most of those predictions match observations very well.

However, predictions for small-scale structure—structure on the scale of dwarf galaxies—are not nearly so reliable. Calculations of the density in the innermost portion of galaxies are too high and predictions for the number of small dwarf galaxies orbiting the Milky Way are too big. Observers don't find the very large number of smaller structures both inside larger halos and in isolated smaller halos that—according to this hierarchical picture—should persist until today.

Perhaps the most well-known discrepancy is known as the *core-cusp problem*. Astronomers and cosmologists predict not only the type of objects that should exist in the Universe, but also how matter should be distributed within them. Predictions for these *density profiles,* as the distributions of mass with distance from objects' centers are known, are *cuspy*. This means that the dark matter density is predicted to peak rather sharply toward the centers, leading to very dense central regions of galaxies and galaxy clusters.

But observers measure density distributions (to some extent) and fail to confirm this prediction. In fact, according to what they have measured so far, most galaxies are not cuspy and exhibit instead what is known as *cored profiles*. (See Figure 37.) This term is confusing, and not only because the Samsung Galaxy Core was the name of a smartphone. Most people probably think of a core as being dense, as is true of the Earth's molten core, for example. A cored density profile refers to the opposite situation—cored in the sense that matter in the center is removed, as when you excise a core from an apple. Of course, no one is removing the entire center of a galaxy. But observations indicate that matter densities aren't peaked nearly as strongly

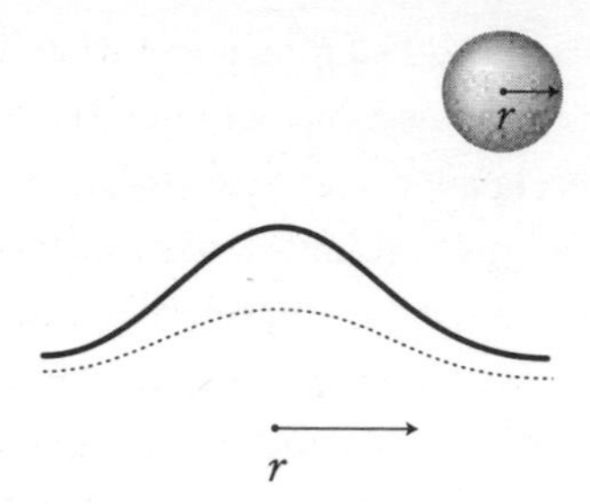

[FIGURE 37] Simulations indicate that in a galaxy, dark matter should have a cuspy density distribution in which matter is heavily concentrated near the center. But observations indicate a cored profile—a smoother, less dense matter distribution in the central region. In both cases, the density peaks in the center as indicated by the figure, but the cuspy distribution peaks far more sharply.

toward the centers of galaxies as predicted. Instead, the central regions' density profiles are relatively flat in these cored regions. The same might be true of galaxy clusters.

Explaining why density profiles look flat or cored and not cuspy, as per dark matter predictions, is an important challenge to the simplest dark matter models. This, along with the *missing satellite problem* (fewer dwarf galaxies than predicted orbiting around bigger central galaxies) and the *too big to fail problem* (a related issue in which the predictions for the densest, most massive galaxies do not agree with observations), possibly point to inadequacies of the standard cold dark matter paradigm.

Another, even more recently noted problem with dark matter predictions—one that the type of dark disk model that I'll soon get to might address—is that the satellite dwarf galaxies found orbiting around larger galaxies don't seem to have the spatial distribution that astronomers would have anticipated. Whereas the expectation was that dwarf satellite galaxies would be distributed more or less spherically in all directions, about half of the approximately thirty dwarf galaxies found orbiting the Andromeda galaxy lie roughly in a plane—and those in the plane almost all have a common orbital direction. A similarly odd distribution is found for some of the dwarf galaxies orbiting the Milky Way.

The close alignment and common rotational direction of the dwarf galaxies might indicate they originated from the disks of merging galaxies. However, even if mergers explain the spatial distribution, the dwarf satellite galaxies problematically seem to contain far too much dark matter for the simplest explanation to be correct. A nonstandard model of dark matter might be called for to explain the dark-matter-dominated dwarf galaxies distributed along a plane.

For all these reported discrepancies, the numerical and observational results are preliminary. At least some of these problems could disappear if our assumptions, observations, or simulations prove to be unreliable. More accurate simulations might demonstrate the imprecision of the initial results or an inadequate understanding of ordinary matter's implications—such as supernovae feedback on structure formation—in which case conventional dark matter models will suffice to explain the Universe's observed structure even without any modifications to dark matter properties. But if the problems persist, the remaining disagreements could turn into real liabilities for the simplest dark matter models and demonstrate the need for more intricate dark matter properties.

When reflecting on these results, it is perhaps heartening to remind ourselves that until well into the 1990s, early simulations—which neglected dark energy—also gave rise to predictions that didn't match data. Many scientists similarly thought then that the early simulations and observations were deceptive and further improvements in data and predictions would reconcile the results. It was a tribute to these early calculations and an indication that accurate predictions can portend new insights that the discrepancies were eliminated once the effect on structure formation of dark energy—an entirely new discovery—was taken into account. Perhaps these current quandaries regarding small-scale structure will similarly be resolved only with a new discovery about the underlying physical properties of matter and energy in the Universe. Observational and computational advances in the next decade will determine if this is indeed the case.

POSSIBLE IMPLICATIONS

Despite the absence of definite confirmation, a number of astrophysicists and cosmologists have started taking the discrepancies seriously and have been investigating the possibility that dark matter does have interactions aside from gravitational. Some go even further and speculate that Einstein's equations of gravity are not quite correct. Despite the attention a few physicists pay to modifications of gravity, I find this more extreme option to be highly unlikely. The evidence for ordinary gravity acting on dark matter that I described earlier on is just too convincing.

Most constraining is the difficulty of explaining observations like that of the Bullet Cluster. This and other similar objects—consisting of galaxy clusters that have merged to leave interacting gas in the middle and dark matter in the outer regions that has simply passed through—would be difficult to explain by anything other than feebly interacting dark matter acting in accordance with the usual gravity equations. In any event, before considering a radical alternative with no consistent theoretical underpinning, we should first consider other, more "boring" reasons that predictions might give misleading results—such as ordinary matter playing a bigger role than simulations have assumed or dark matter being different than determined by conventional expectations.

I recently attended two conferences where issues with small-scale structure and their possible resolutions were discussed. The first was a small workshop that my particle physics colleagues at the Harvard physics department had organized on the topic of self-interacting dark matter. Astrophysicists from the Harvard Center for Astrophysics organized the second conference, titled "Dark Matter Debates," which took place the following spring in 2014. Fortunately the debates were about substance and not opinions—which when overemphasized can derail many a scientific discussion.

One reason I found these conferences so worthwhile was the ample opportunity they provided for conversations among physics and astronomy colleagues at Harvard. The Center for Astrophysics, where the astronomers work, was built in 1847 on the highest point in Cambridge to house the largest telescope in the United States of its time—a paltry fifteen inches in diameter. The location—which persists even though the telescope's scientific standing does not—keeps astronomers and physicists a mile apart, so we never accidentally meet each other at the watercooler or the coffee machine. The conferences brought us—and many visiting physicists and astronomers—to the same place.

But the primary merit of the conferences was that the results that were presented were original and very new. The topics included existing evidence for small-scale structure problems and the possible resolutions, which conference participants argued could be either an inadequate appreciation of the role of ordinary matter or something truly novel like dark matter self-interactions.

The speakers who discussed why ordinary matter could influence the formation of structure on small scales argued that the inclusion of ordinary matter in numerical simulations—the most prosaic possibility—might go a long way toward resolving the discrepancies between predictions and observations. Initial simulations assumed that dark matter dominates the dynamics and the growth of structure, whereas ordinary matter simply falls into the gravitational potential wells that dark matter creates. Though it might light up the more massive, denser regions after stars have formed, ordinary matter's influence was neglected apart from this role as a spotlight on dense dark matter locations.

Physicists initially thought that ordinary matter wouldn't significantly affect the growth of structure and furthermore was too computationally imposing to reliably include anyway. Even today, when astronomers do try to include the effects of ordinary matter, a good deal of uncertainty remains. With existing memory and computing power, no one can simulate everything in detail, so astronomers doing simulations need to use approximations and assumptions. Even so, the limited numerical simulations currently underway that take account of ordinary matter do seem to alleviate the discrepancies.

A couple of effects account for the improved agreement between simulations and observations. Standard matter interacts through forces aside from gravity, so even though its initial gravitational influence is relatively small, its impact on structure—especially on smaller scales—might not be. For example, one potential explanation for the paucity of observed satellites is that they are just too faint. Intergalactic gas can be heated when ultraviolet radiation—stuff relating to ordinary matter—is emitted from stars. Once this happens, halos—particularly small ones—will be less effective at accreting gas. But if halos don't have enough gas, they won't form stars, which could leave them too dim to observe with current telescopes.

Another proposed explanation for the paucity of observed satellite galaxies and sparser-than-expected inner galaxy cores is that supernova explosions expel material out of the inner portions of their host galaxies, leaving behind a far less dense inner core. The resulting dark matter distribution might be compared to that of an urban population in its densest inner city regions, where—in the aftermath of unrest—explosions of violence have stemmed the growth to leave a depleted core. The inner galaxy that has seen too much supernova outflow doesn't grow in density toward the center any more than would a sparsely occupied inner city.

Furthermore, the energy released by supernova explosions can ionize and heat the gas in the outer regions of a galaxy. This can blow away a lot of the ordinary matter that might have condensed into dwarf satellite galaxies that rotate around a bigger one or hinder the collapse into dense regions that are required for star formation. These outer dwarf galaxies would have proportionally less ordinary matter and would be fainter, making them far more difficult to find.

The evidence for and against a big role for ordinary matter in small-scale structure is constantly evolving as methods and computing power consistently improve. Several animated discussions broke out at the conferences—though as noted by a particle physics colleague, the tone among the astronomers was pleasantly conciliatory, with everyone trying to find the right answers and not just beating home a point. Even those who emphasized the importance of ordinary matter acknowledged that it won't suffice to eliminate all discrepancies if the smallest-scale problems persist in isolated dwarf galaxies, where supernova feedback is expected to be very weak. If that is the case, as current observations suggest, something beyond ordinary dark matter will be called for. Although everyone at the conferences agreed that ordinary matter can go in the right direction to help resolve disagreements between simulations and data, the physicists and astronomers who were present all recognized that incorrect predictions for satellite galaxies might well suggest a more radical modification of the standard non-interacting dark matter paradigm.

SELF-INTERACTING DARK MATTER

Given the intriguing issues that have arisen when simulations have confronted data, it is interesting to consider alternative dark matter models that might address them. The most intriguing possibility is that the assumption of simple non-interacting dark matter is misguided and dark matter interactions apart from gravitational influence structure. Considering this possibility helps physicists learn more about dark matter particles' interactions among themselves, and what new forces might act on them. Ongoing measurements and improving simulations will tell us more about the nature of dark matter—no matter what the results turn out to be. Even if discrepancies don't survive, we will better understand the nature of dark matter and how dark matter and ordinary matter contribute to cosmic structure. But if disagreements persist, they might well be evidence for self-interactions.

Self-interacting dark matter is a promising new suggestion, in part because we know so little about dark matter's properties. Just as ordinary matter can experience nongravitational forces such as electromagnetism, so too might dark matter. Although the usual assumption is that that dark matter experiences gravitational interactions and possibly very feeble interactions with ordinary matter, our not having directly detected dark matter in experiments tells us nothing about dark matter's interactions with itself. Self-interacting dark matter particles would attract or repel other dark matter particles, but not those of the matter with which we are familiar. Dark matter could experience as-yet undetected dark forces, which would influence dark matter particles but not those of ordinary matter. Because standard forces such as electromagnetism act only on ordinary matter and dark forces would act only on dark matter, dark matter and ordinary matter particles would remain essentially oblivious to each other.

Self-interacting dark matter, like ordinary matter, would be social. But it would be cliquish too—interacting only with its own kind. Dark matter might scatter off other dark matter particles, but ordinary matter would be as invisible to it as dark matter is to ordinary stuff. Since direct detection experiments look only for interactions between dark matter and ordinary matter, such a possibility is not ruled out, and it might even be favored by structure investigations.

We don't know what form the new forces will take if dark matter does interact with itself. Even so, forces among dark matter particles are constrained and dark matter self-interactions cannot be too strong. Recall that the "smoking gun" signature of dark matter is the Bullet Cluster formed from merging clusters, as well as other galaxy clusters that take a similar form. Gravitational lensing observations tell us that the dark matter from one cluster passed through dark matter from another one essentially unimpeded—leading to two bulbous shapes in the outside regions and gas that remains trapped in the central region in between.

If all dark matter very strongly interacted with itself—as ordinary matter does—it would act like the gas and remain in the center. But the bulbous outer regions indicate that dark matter didn't do that and instead passed right through. This doesn't tell us that dark matter can't interact at all. But it does limit the strength and distance scale over which the force can be relevant. The strength of dark matter interactions is constrained too by the shapes of galaxy haloes, which are also sensitive to dark matter interactions.

But these restrictions don't rule out self-interactions as a possibility. They just provide limits on the allowed strength and form. Even allowing for these constraints, self-interactions can in principle be strong enough to address the problems with small-scale structure predictions. Speakers at the conferences showed why self-interactions of dark matter can help with some of the potential structure issues and furthermore lower the core densities of the most massive satellites, yielding better agreement between predictions and observations.

For example, self-interactions could address the problem of too much predicted density in the central region of galaxies. In the absence of nongravitational interactions, dark matter would keep falling into the center since dark matter that moves sufficiently slowly gets trapped in the gravitational potential of existing structures, making the density at the center rise dramatically. But repulsive interaction among dark matter particles would keep dark matter particles apart—preventing them from piling in too closely. It would be as if everyone in a packed train station surrounded themselves with their luggage, keeping everyone else at arm's distance. Repulsive interactions among dark matter particles would similarly introduce a protective barrier—keeping dark matter from becoming too dense.

Simulations with self-interacting dark matter confirm this intuition, and indeed lead to cored halo shapes—those with relatively constant-density inner regions—rather than cuspy ones. The matter density toward the center of a galaxy or galaxy cluster can only increase so much before it saturates and can't become any denser. Any persistent small-scale structure issues could potentially be resolved if dark matter interacts in this way.

The different predictions of self-interacting dark matter for galaxies and for galaxy clusters guarantee that future observations and simulations will tell us a good deal more about dark matter's properties. And, since different models for the interactions make different predictions, comparing simulations to observations should furthermore distinguish among different types of interactions.

Only by considering alternatives can we exploit the abundant data about the shape of structures in the Universe in order to more fully understand their implications. Maybe dark matter has interactions that affect structure so that simulations will better agree with observations. Or maybe ordinary dark matter does a better job in predicting structure than dark matter with such interactions, allowing us to decisively rule out more elaborate models once the simulations and measurements are truly reliable. Whatever the result, we will learn a lot beyond what the conventional WIMP searches described in the previous chapter will tell us.

But interacting dark matter in itself—interesting as it might be—is not precisely the focus of my recent research, which the next chapter will present. After all, interacting and non-interacting dark matter are not the only possibilities. Just as by focusing on black and white, we neglect shades of gray—not to mention polka dots and stripes—assuming that dark matter either all has interactions or that none of it does leads us to overlook the richness of the world's possibilities. The next chapter will consider the intriguing notion that dark matter—like ordinary matter—is indeed more complex. Perhaps dark matter has a non-interacting component and a self-interacting one too, and both of them contribute to the Universe's structure and behavior.

19

THE SPEED OF DARK

Both casual observers of science and scientists themselves frequently employ Occam's Razor for guidance when evaluating scientific proposals. This oft-cited principle says that the simplest theory that explains a phenomenon is most likely to be the best one. Its sensible-sounding logic dictates that it is probably a bad idea to build complicated structures when a leaner one will do.

Yet two factors undermine the authority of Occam's Razor, or at least suggest caution when using it as a crutch. I've learned the hard way to be wary of crutches—both intellectual and physical. Once when I used the tangible kind to heal a broken ankle, I leaned on them incorrectly and caused nerve damage in my arms. Theories that conform to the dictates of Occam's Razor sometimes similarly address one outstanding problem while creating issues elsewhere—usually in some other aspect of the theory that embraces it.

The best science should always encompass, or at the very least be consistent with, the broadest possible range of observations. The real question is what most effectively solves the entire set of unexplained phenomena. A proposed explanation that seems simple at first might devolve into a Rube Goldberg contraption when confronted with a larger set of issues. On the other hand, an explanation that seemed unduly cumbersome when applied to the original problem might, when seen through the lens of scientific peripheral vision, come to reveal its underlying elegance.

My second concern about Occam's Razor is just a matter of fact. The world is more complicated than any of us would have been likely to conceive. Some particles and properties don't seem necessary to any physical processes that matter—at least according to what we've deduced so far. Yet they exist. Sometimes the simplest model just isn't the correct one.

Discussions on this topic emerged many times at the "Dark Matter Debates" conference mentioned in the previous chapter. In her talk about experimental constraints on unnecessary but testable particles, the particle physicist Natalia Toro argued that a more appropriate guide than Occam's Razor would be a principle she called "Wilson's Scalpel." She named it after the physicist Ken Wilson, who developed a general framework for understanding how to do science by keeping track only of those elements that are testable. Natalia proposed that a scalpel in his name should be used to shape, rather than shave, a theory, leaving all testable elements intact—whether or not we can attribute an underlying purpose. When I spoke next, I jokingly suggested that the principle of "Martha's Table" was a better idea still. After all, you don't set a table with just knives. You set it with everything necessary to eat a meal in a dignified fashion. With the talent of a Martha Stewart, you will retain an organizing principle—no matter how many dishes and silverware you arrange.

Science too should have a properly set table—one that allows us to address the many phenomena that we observe. Although scientists tend to prefer simple ideas, they are rarely the whole story.

This discussion is all by way of prelude to my introducing what my collaborators and I called "partially interacting dark matter," which led to the "double-disk dark matter" category of models, which I will also now present. Both classes of models acknowledge that the makeup of dark matter might not be so simple. Just as is true of particles of ordinary matter, dark matter particles might not all be the same. New types of dark matter with different types of interactions might exist and furthermore have observable, previously unforeseen consequences. Even if the interacting component turns out to be only a small fraction of the dark matter, it could have important implications for the Solar System and the galaxy. It might have had some significance for the dinosaurs too.

ORDINARY-MATTER CHAUVINISTS

Even though we know that ordinary matter accounts for only about one-twentieth of the Universe's energy and a sixth of the total energy carried by matter (with dark energy constituting the remaining portion), we nonetheless consider ordinary matter to be the truly important constituent. With the exception of cosmologists, almost everyone's attention is focused on the ordinary matter component, which you might have thought to be largely insignificant according to the energy accounting.

We of course care more about ordinary matter because we are made of the stuff—as is the tangible world in which we live. But we also pay attention because of the richness of its interactions. Ordinary matter interacts through the electromagnetic, the weak, and the strong nuclear forces—helping the visible matter of our world to form complex, dense systems. Not only stars, but also rocks, oceans, plants, and animals owe their very existence to the nongravitational forces of nature through which ordinary matter interacts. Just as a beer's small-percentage alcohol content affects carousers far more than the rest of the drink, ordinary matter, though carrying a small percentage of the energy density, influences itself and its surroundings much more noticeably than something that just passes through.

Familiar visible matter can be thought of as the privileged percent—actually more like 15 percent—of matter. In business and politics, the interacting one percent dominates decision making and policy, while the remaining 99 percent of the population provides less widely acknowledged infrastructure and support—maintaining buildings, keeping cities operational, and getting food to people's tables. Similarly, ordinary matter dominates almost everything we notice, whereas dark matter, in its abundance and ubiquity, helped create clusters and galaxies and facilitated star formation, but has only limited influence on our immediate surroundings today.

For nearby structure, ordinary matter is in charge. It is responsible for the motion of our bodies, the energy sources that drive our economy, the computer screen or paper on which you are reading this book, and basically anything else you can think of or care about. If something has measurable interactions, it is worth paying attention to, as it will have far more immediate effects on whatever is around.

In the usual scenario, dark matter lacks this type of interesting influence and structure. The common assumption is that dark matter is the "glue" that holds together galaxies and galaxy clusters, but resides only in amorphous clouds around them. But what if this assumption isn't true and it is only our prejudice—and ignorance, which is after all the root of most prejudice—that led us down this potentially misleading path? What if, like ordinary matter, a fraction of the dark matter interacted too?

The Standard Model contains six types of quarks, three types of charged leptons including the electron, three species of neutrinos, all the particles responsible for forces, as well as the newly discovered Higgs boson. What if the world of dark matter—if not equally rich—is reasonably wealthy too? In this case, most dark matter interacts only negligibly, but a small component of dark matter would interact under forces reminiscent of those in ordinary matter. The rich and complex structure of the Standard Model's particles and forces gives rise to many of the world's interesting phenomena. If dark matter has an interacting component, this fraction might be influential too.

If we were creatures made of dark matter, we would be very wrong to assume that the particles in our ordinary matter sector were all of the same type. Perhaps we ordinary matter people are making a similar mistake. Given the complexity of the Standard Model of particle physics, which describes the most basic components of matter we know of, it seems very odd to assume that all of dark matter is composed of only one type of particle. Why not suppose instead that some fraction of the dark matter experiences its own forces?

In that case, just as ordinary matter consists of different types of particles and these fundamental building blocks interact through different combinations of charges, dark matter would also have different building blocks—and at least one of those distinct new particle types would experience nongravitational interactions. Neutrinos in the Standard Model don't interact under the strong or electric force yet the six types of quarks do. In a similar fashion, maybe one type of dark matter particle experiences feeble or no interactions aside from gravity, but a fraction of it—perhaps five percent—does. Based on what we've seen in the world of ordinary matter, perhaps this scenario is even more likely than the usual assumption of a single very feebly or non-interacting dark matter particle.

People in foreign relations make a mistake when they lump together another country's cultures—assuming they don't exhibit the diversity of societies that is evident in our own. Just as a good negotiator doesn't assume the primacy of one sector of society over another when attempting to place the different cultures on equal footing, an unbiased scientist shouldn't assume that dark matter isn't as interesting as ordinary matter and necessarily lacks a diversity of matter similar to our own.

The science writer Corey Powell, when reporting on our research in *Discover* magazine, started his piece by announcing that he was a "light-matter chauvinist"—and pointing out that virtually everyone else is too. By this he meant that we view the type of matter we are familiar with as by far the most significant and therefore the most complex and interesting. It's the type of belief that you might have thought was upended by the Copernican Revolution. Yet most people persist in assuming that their perspective and their conviction of our importance are in keeping with the external world.

Ordinary matter's many components have different interactions and contribute to the world in different ways. So too might dark matter have different particles with different behaviors that might influence the Universe's structure in a measurable fashion.

THE INTERACTING MINORITY

My collaborators and I called the scenario with a small component of dark matter that interacts through nongravitational forces "partially interacting dark matter." We first investigated the simplest such model, which involves only two components. The dominant component interacts only gravitationally and is the conventional cold dark matter that resides in spherical haloes around galaxies and galaxy clusters. The second component interacts gravitationally too, but also through an additional force very similar to electromagnetism.

This two-tiered dark matter scenario might sound exotic, but keep in mind that similar statements can be made about ordinary matter. Quarks experience the strong nuclear force but particles like electrons don't. That is why quarks get bound into protons and neutrons but electrons do not. Similarly, electrons experience the electromagnetic force to which neutrinos are oblivious. So if we go against our usual chauvinism and allow for similar diversity in the dark world, it shouldn't be impossible to imagine that a portion of the dark sector interacts through forces similar to—but distinct from—the ones through which the stuff we are made of interacts.

However, bear in mind that partially interacting dark matter is a little different from Standard Model matter in that although electrons don't experience the strong force directly, they do interact with quarks and therefore experience indirect effects. The newly proposed form of dark matter might instead be entirely isolated in its interactions, with the bulk of dark matter not even experiencing indirect effects of the newly introduced dark force. Since we don't yet know whether components of dark matter should interact—or whether dark matter is even composed of different types of particles—the first and simplest assumption would be that there are no other new interactions aside from the new form of electromagnetism and only the newly introduced charged particles experience this force. In this scenario, the bulk of the dark matter wouldn't experience the new force at all.

For fun, I'll call the force that is experienced by the interacting dark matter component *dark light*, or more generally I'll call it *dark electro-*

magnetism. The names are chosen to remind us that the new type of dark matter experiences a force like electromagnetism—but one that is invisible to the ordinary matter of our world. Whereas ordinary matter carries charge so that it emits and absorbs photons, the newly introduced component of dark matter would emit and absorb only this new type of light, which ordinary matter simply doesn't experience.

This dark electromagnetic force would be analogous to the usual electromagnetic force. But it would be an entirely different influence acting on particles charged under a distinct additional force that is communicated by an entirely new type of particle—a dark photon if you will. Though the new component of dark matter wouldn't interact with ordinary matter, it would have self-interactions that would make it behave similarly to familiar matter, which, after all, doesn't interact with dark matter either.

Both ordinary matter and dark matter would carry charge and experience forces, but those charges and forces would be distinct. The particles that carry charge under the new dark force would be attracted or repelled by each other in a way that is similar to the behavior of ordinary charged particles. But the dark sector's interactions would be transparent to ordinary matter since dark matter interacts through its own unique form of light—not through the light with which we are familiar. Only dark matter particles would experience the new force's influence.

Even while obeying similar laws of physics and maybe being in close proximity in space, dark matter and normal matter would each occupy their own worlds. Ordinary matter and dark matter could even physically overlap without ever interacting. Because they would interact with each other through distinct forces—aside from their extremely feeble gravitational influence, the charged ordinary matter and charged dark matter would be oblivious to each other's presence.

Two types of electrically charged particles in the same place that don't interact with each other is really not so mysterious. It's a bit like ordinary matter interacts via Facebook whereas the charged matter of the partially interacting dark matter model interacts on Google+. Their interactions are similar, but they have contact only with those on their own social network. Interactions proceed on one network or the other—but usually not on both.

Going further afield to make an analogy, it's like left-wing and right-wing TV shows, which follow more or less the same rules of programming and can both be broadcast on a single television, but which are entirely different entities—each reinforcing their own confirmation biases. Though they have similar formats, with interview hosts, guest "experts," graphic displays illustrating their points, and running news alerts on random unrelated topics underneath, the actual content and outcomes as well as the advertisers for the two types of shows are nonetheless very different. Very few if any guests or issues will appear on both types of show and the products and candidates they advocate will be different too.

Just as it's only rare individuals who both watch Fox News and listen to NPR, most, or perhaps all, particles interact via one force or the other. The model—like the media—encourages sticking to one point of view. Though in principle there can be intermediary particles that interact via forces of both types, most particles carry either one type of charge or the other and therefore don't communicate with each other.

To be fair, it wasn't only prejudice that discouraged physicists from thinking about a new type of electromagnetism that dark matter would experience. Interactions have consequences that often can be tested. Physicists shied away from the idea of dark forces and self-interacting dark matter because they thought such scenarios were constrained or even ruled out. However, as explained in Chapter 18, even if all of dark matter experiences those forces, those constraints are not so severe. But interactions are allowed only within prescribed limits based on observations.

However, the situation is much less constrained if only a small portion of the dark matter self-interacts. Recall the two types of limits on self-interactions. The first had to do with the structure of halos themselves: they had to be spherical—with a little non-uniformity known as triaxial structure. The second concerned galaxy cluster mergers, such as the most famous one, known as the Bullet Cluster, which was the result of merging clusters. The gas visibly remains in the central region, but dark matter, observed through gravitational lensing, passed through unhindered to create two external bulbous structures—a bit like Mickey Mouse's ears.

Both constraints are most significant when all of the dark matter interacts. But neither tells us much if the interacting component constitutes only a small fraction of the dark matter. If only a small component interacts, most of the halo will be spherical. The interactions won't wipe out the triaxial structure either unless it is the dominant component or scatters more than can be expected.

Similarly, the fractions of gas and dark matter in the Bullet Cluster are not nearly sufficiently well measured to register a tiny component of dark matter, which, after all, comprises only a small fraction of the galaxy cluster. That component might interact and remain in the central region along with the gas—and no one would be any the wiser for it. Perhaps eventually measurements such as those of the Bullet Cluster will become sufficiently precise to constrain the partially interacting scenario I'm describing. Certainly now, partially interacting dark matter remains a viable and promising possibility.

THE SPARK

My impetus for considering this idea—along with Matthew Reece, a recent young addition to the Harvard physics faculty, and two postdoctoral fellows, JiJi Fan and Andrey Katz—was not entirely direct. As with many other research projects that turn out to be the most interesting, we weren't aiming to study what ultimately became our major focus. Rather we were trying to understand some intriguing data from the Fermi satellite—the NASA space observatory that scans the sky for gamma rays, which are a more energetic version of light than visible light or even X-rays.

Most astrophysical processes produce radiation with a smooth distribution over a broad range of frequencies, meaning the number of photons doesn't change dramatically at any particular wavelength. So when Christoph Weniger from the University of Amsterdam and his collaborators noticed an excess of radiation in the Fermi data all concentrated at a single frequency, it sparked our interest—and that of many others in the physics and astronomy communities.

The spike in the density of radiation (here radiation just means photons or light) that Weniger and his collaborators had identified appeared to emerge from the center of the galaxy, where dark matter is more highly concentrated, but where no such signal from ordinary astrophysical sources should arise. In the absence of a more conventional explanation—or a mistake—a spike in the photon number could only represent something new.

The most intriguing suggestion was that the signal could be the result of dark matter annihilating into photons—an indirect detection signal of the sort that was described in Chapter 17. Maybe dark matter particles collided with each other and through the "magic" of $E = mc^2$, turned into photons that the Fermi satellite could then detect. Further support for this suggestion was that the energy of the photons at which the excess was observed was in the range expected for dark matter. It was also close in value to that of the Higgs boson mass—the mass of the recently discovered missing piece of the particle physics Standard Model—perhaps indicating an even deeper connection. The third intriguing aspect of the measurement was that the interaction rate agreed with what it takes to get the right dark matter relic density. Just about the right amount of dark matter would remain today if dark matter annihilated at the rate that had been measured.

Despite these rosy signs, however, a few things seemed badly off if indeed the signal originated in dark matter. Dark matter doesn't produce photons directly since it doesn't interact with light. Maybe dark matter interacts with some heavy charged particle that we haven't yet observed and those particles in turn interact with light. But if that were the case, we would have expected that when dark matter annihilates and turns into energy, that energy would produce charged particles too. But the Fermi satellite detected no sign of such a process.

The other problem was that although the total amount of dark matter depends on how much it annihilates to anything, the signal depends only on the amount it annihilates to photons. Given the dark matter density in the Universe, the annihilation rate to photons turned out to be too small in all but the most finely-tuned models. This meant that this particular dark matter explanation of the signal could only even possibly be consistent in a very narrow range of parameters that would permit a large enough rate of annihilation to photons but no measurable annihilation into charged particles. No credible scenario seemed to make this happen.

JiJi, Andrey, Matt, and I viewed this as an interesting opportunity to explore the range of permissible dark matter models. We wanted to know if there was any reasonable example in which all the rates agree with their measured values. We began by focusing on the Fermi result and asking whether we could think of a way that nature could do better than the models other physicists had already suggested. We were fully aware that the data might turn out to have been misleading. The Fermi results were tantalizing, but were not sufficiently strong to make a decisive case for a new signal—with an origin in dark matter or otherwise. The observations might have simply reflected some statistical fluke or a misunderstanding of the apparatus, rather than a true signal of a new physical process, which—to stem any overly high expectations as you read this—turned out to be the case.

But the observation was sufficiently interesting that, especially early on, it warranted asking whether any reasonable physical process could have created it. After all, looking for exotic new forms of matter is tough. We want to be aware of every possible way of finding them. Whether or not this signal would prove to be correct, we might learn something that could be useful in the future.

The four of us worked at my blackboard, trying out a number of ideas designed to cleverly escape the problems while preserving the desirable features of the signal. But none of our proposals worked well enough to be worth pursuing. The ones that succeeded in satisfying all the constraints were inconsistent with the spirit of Occam's Razor. Worse still, they wouldn't be allowed anywhere in the vicinity of a properly set table of ideas.

However, one of the models that we rejected triggered a line of thinking that was ultimately much more interesting than anything we had set out to do. Our initial inquiries were all based on trying to find a particular model that we could shoehorn into existing constraints. But we took a step back and asked ourselves: What if the local dark matter was denser than we thought so that we were in fact misinterpreting the implications? What if dark matter could annihilate much more than expected because of this greater density?

With higher density, dark matter particles could find and interact with each other much more efficiently. This in turn would create a bigger signal that would more readily stand up to observations. Just as you are more likely to bump into someone in New York's Penn Station at rush hour than you are in the Waterbury, Vermont train station at 9 A.M. on a Sunday, one dark matter particle is more likely to interact with another dark matter particle in a dense matter environment than in the usual diffuse environment of the amorphous halo. If some dark matter were more concentrated than the matter in the halo, all the other constraints could be much more readily satisfied.

The question then is the underlying reason. Why would dark matter—or at least some of the dark matter—be denser than we thought? This is where the idea of partially interacting dark matter came about—together with the dark-disk idea that closely follows. In fact, even though we now are pretty sure the Fermi signal is spurious, this new idea has so many unexplored implications that we soon realized it was independently worth pursuing. One of these consequences is a disk of dark matter with far greater density than is usually assumed.

THE DARK DISK

Once when I was cleaning my house (well, letting my Roomba robotic vacuum cleaner do its thing), I emptied the dust tray and found an old fortune cookie paper I had saved. The paper asked an enigmatic question, "What is the speed of dark?" I didn't know then that the words were indeed a sort of fortune, in that they more or less prophesied the research project I was about to commence.

Chapter 5 explained that ordinary matter is found in a thin dense disk because it sheds energy, which it does by emitting photons that efficiently carry energy away. The consequence of dissipation of energy is slower, cooler matter particles that don't make the kind of big excursions that would be expected from hotter, more energetic, higher-velocity ones. Matter collapse happens because with less energy, it has less velocity with which to spread out. Ordinary matter, which dissipates energy and thereby lowers its speed, collapses into a disk—like the disk of the Milky Way—which you can see on a clear, dry night.

After my collaborators and I had unleashed the notion of partially interacting dark matter, we pursued its potential consequences for the Milky Way galaxy and beyond. We assumed that interacting dark matter is present, and that it behaves similarly to ordinary charged matter, which we know in the galaxy cools, slows down, and thereby forms a disk.

Only a small fraction of the dark matter interacts in our scenario. So the bulk of the dark matter would still form a spherical halo, consistent with what astronomers have so far observed. However, the new component of interacting dark matter could dissipate energy so it—like ordinary matter—could cool down and form a disk too. The interacting dark matter component would—via dark photon interactions—radiate away energy and lower its velocity. In this respect it would behave very much like ordinary matter. Just as ordinary matter cools and collapses, the interacting component of dark matter would as well. And because of conserved angular momentum, which prevents collapse in all but the vertical direction, the interacting dark matter would collapse into a disk.

Furthermore, just as ordinary atoms are composed of protons and electrons with opposite charges, this component of dark matter would also contain oppositely charged particles. The charged particles would continue to radiate energy until they became cool enough to get bound into dark atoms. The cooling would then become much slower and the dark matter atoms, like the atoms of normal matter, would reside in a disk whose thickness would be related to the temperature at which atomic binding occurred. With reasonable assumptions, the temperatures of the ordinary and dark components after cooling stops should turn out to be comparable. So we would be left with a disk of dark matter and a disk of ordinary matter whose temperatures would be about the same.

However, the dark disk wouldn't have exactly the same structure as the usual Milky Way disk. In fact, it might even be more interesting. The remarkable property of the dark disk is that if a dark matter particle is heavier than a proton but has the same temperature, the dark disk will be thinner—narrower in width than that of the Milky Way. The energy a particle carries is associated with its temperature. But kinetic energy is also related to mass and velocity. Heavier particles with the same temperature will have lower velocity in order for the energy to be about the same, so bigger masses lead to thinner disks. For a dark matter particle with mass about one hundred times heavier than the proton—a commonly assumed value for dark matter masses—the disk could be about one hundred times thinner even than the narrow disk of the Milky Way—a remarkable possibility, which, as we will see in the next couple of chapters, can give rise to many interesting observational consequences. (See Figure 38.)

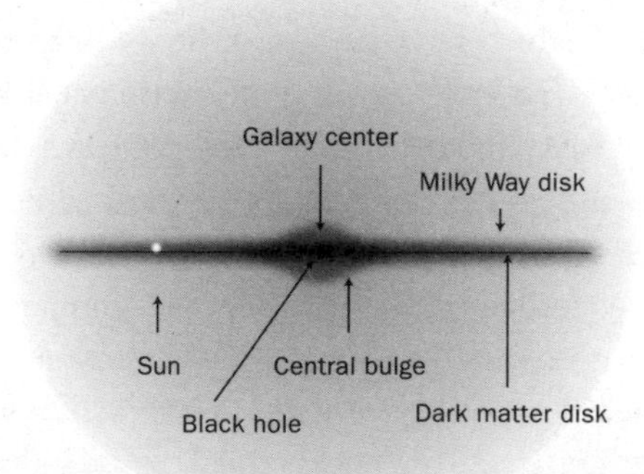

[FIGURE 38] A small interacting component of dark matter can lead to a very thin dark disk in the midplane of the Milky Way, indicated in the picture by the solid black line.

Also important is that the two disks, though different, should nonetheless be aligned—with the dark disk embedded inside the wider disk of the Milky Way plane. That's because the ordinary matter and dark matter disks, which interact via gravity, are not entirely independent. Reflecting one flaw in my earlier Fox News–NPR analogy, the gravitational pull that both the dark disk and the ordinary disk experience would actually make the two entities want to orient themselves in the same direction. Though left- and right-wing TV are not entirely independent either since they do influence each other through the collective effects of their constant and often repeated broadcasts, most reactions are negative, which makes their mutual interaction repulsive. The dark matter and normal matter disks, on the other hand, interact via gravity and therefore align.

The remarkable and surprising outcome of our research was that there can be a thin disk of dark matter that exists along with our ordinary disk—and that this newly proposed dark matter disk can be embedded inside the well-known one of the Milky Way. My collaborators and I were rather excited about our proposal and were eager to share it with other physicists. My colleague Howard Georgi at Harvard very much liked the idea too, but wisely thought this scenario merited a catchier name than anything we had suggested. He did us the further favor of proposing the alternative name "double disk dark matter" (DDDM), which served our purpose well and which we have since employed. The name is appropriate since according to our assumptions, the galaxy indeed contains two types of disks, with one embedded inside the other.

Star observations indicate that we left the center of the plane less than a couple of million years ago—a short time on cosmological scales. This tells us that if double-disk dark matter exists, the Solar System oscillated through the dark disk around that time too, so we aren't very far away (in astrophysical terms). In fact, if the disk turns out to be a little thicker, we might even be inside it—perhaps with observable consequences. And, as we will soon see, the disk would also influence the dynamics of the Solar System—possibly with some very dramatic effects, albeit on very long time scales. The small component of interacting dark matter that we proposed can also create disks inside other galaxies—possibly explaining some of their properties too.

Of course, the big question is whether an interacting component of dark matter and dark matter disks truly exist. Discovering a dark disk by measuring its consequences would help establish the significance of any of the above suggestions. Fortunately, as with ordinary matter, even if it is only a small fraction of the total dark matter in the Universe, the interacting component's enhanced density might make it easier to find and identify than the usual diffuse dark matter in the halo. The many potential particle physics and astronomical signals of this enhanced dark matter density that the next chapter will discuss should tell us if the dark disk is viable or perhaps even preferred.

If I am truly lucky—as the fortune hiding in my Roomba might have me believe—perhaps one or more of these observations will ultimately reveal a dark matter disk's existence.

20

SEARCHING FOR THE DARK DISK

I recently participated in a spirited discussion with lawyers, academics, writers, and human rights workers on the topic of free speech. None of us questioned free speech's importance. However, we didn't all agree on what exactly it should mean, or how we should balance it in the context of other rights. When does free speech's potentially harmful consequences outweigh its benefits? Should spending money to promote particular laws or candidates be limited in any ways? A lawyer explained how the U.S. Supreme Court relied on the right to free speech—along with the idea of spending money as a form of expression—to decide the *Citizens United* case, which allows unrestricted political contributions from corporations. But others in our discussion were concerned that unlimited spending by corporations can drown out individual citizens' voices—arguing too that free speech was intended for people, not corporations. After all, neither money nor corporations can speak—freely or not—without a human voice.

But that Supreme Court decision being what it is, and with the resulting flood of money that now enters politics, let's consider the different ways that individuals—and corporations—can spend their money to influence public opinion.

Financial contributions might be focused to target advertising in local regions such as cities and towns, where they can readily change people's viewpoints and influence the outcome of a vote. Or donors can contribute more diffusely, spreading their wealth and blanketing their claims over a larger region—yielding some general shaping of opinion but generating less clearly delineated effects. The two strategies together hold stronger sway than either type of advertising alone. But the rate of change should be greater in the targeted regions, clearly reflecting the greater density of advertising in the smaller but more concentrated locales.

Similarly in physics, the gravitational influence of a thinner, denser disk would more sharply influence the motion of stars than a thicker, more diffuse one. As with the more pronounced influence of local advertising, the positions and velocities of stars moving in and out of the galactic plane would be more noticeably influenced by a thinner, denser disk.

Because the Milky Way would contain disks of both ordinary matter and dark matter, the stars' motions in and out of the galactic plane would depend on both, generating a combined influence that varies sharply, then gradually, as you move away from the dense region at the galactic midplane—analogous to the consequences of local and global advertising together. With a thin dark disk embedded in a thicker ordinary matter one, dark matter's concentrated pull would combine with the more diffuse tug of ordinary matter to yield a distinctive measurable influence on stars that would vary with the distance from the midplane of the Milky Way.

We live in a data-rich era and we certainly don't want to overlook any possible search targets—especially when looking for something as astounding but elusive as a disk of dark matter. This chapter will explain how measuring the Milky Way disk's gravitational influence using the motion of stars will establish or undermine a dark disk's existence. But before getting to that, this chapter will first explore other general considerations about disk dark matter's possibilities and the potential for its discovery through more conventional dark matter searches that are currently under way. After that, it will present some of the dark disk's intriguing astronomical implications.

DIVERSE DARK MATTER

When first studying partially interacting dark matter, I was astonished to find that practically no one had considered the potential fallacy—and hubris—of assuming that only ordinary matter exhibits a diversity of particle types and interactions. Although a few physicists had tried to analyze models such as one known as *mirror dark matter*, which features dark matter that mimics everything about ordinary matter, exemplars such as this one were rather specific and exotic. Their implications were difficult to reconcile with everything we know.

A small community of physicists had studied more general models of interacting dark matter. But even they assumed that all the dark matter was the same and therefore experienced identical forces. No one had allowed for the very simple possibility that although most dark matter doesn't interact, a small fraction of it might.

One potential reason might be apparent. Most people would expect a new type of dark matter to be irrelevant to most measurable phenomena if the extra component constitutes only a small fraction of the dark matter inventory. Having not even observed the dominant component of dark matter, concerning oneself with a smaller constituent might seem premature.

But when you remember that ordinary matter carries only about 20 percent of the energy of dark matter—yet it's essentially all that most of us pay attention to—you can see where this logic could be flawed. Matter interacting via stronger nongravitational forces can be more interesting and more influential even than a larger amount of feebly interacting matter.

We've seen that this is true for ordinary matter. Ordinary matter is unduly influential given its meager abundance because it collapses into a dense matter disk where stars, planets, the Earth, and even life could form. A charged dark matter component—though not necessarily quite as bountiful—can collapse to form disks like the visible one in the Milky Way too. It might even fragment into starlike objects. This new disklike structure can in principle be observed, and might even prove to be more accessible than the conventional dominant cold dark matter component that is spread more diffusely in an enormous spherical halo.

Once you start thinking along these lines, the possibilities quickly multiply. After all, electromagnetism is only one of several nongravitational forces experienced by Standard Model particles. In addition to the force that binds electrons to nuclei, the Standard Model particles of our world interact via the weak and strong nuclear forces. Still more forces might be present in the world of ordinary matter, but they would have to be extremely weak at accessible energies since so far, no one has observed any sign of them. But even the presence of three nongravitational forces suggests that the interacting dark sector too might experience nongravitational forces other than just dark electromagnetism.

Perhaps nuclear-type forces act on dark particles in addition to the electromagnetic-type one. In this even richer scenario, dark stars could form that undergo nuclear burning to create structures that behave even more similarly to ordinary matter than the dark matter I have so far described. In that case, the dark disk could be populated by dark stars surrounded by dark planets made up of dark atoms. Double-disk dark matter might then have all of the same complexity of ordinary matter.

Partially interacting dark matter certainly makes for fertile ground for speculation and encourages us to consider possibilities we otherwise might not have. Writers and moviegoers especially would find a scenario with such additional forces and consequences in the dark sector very enticing. They would probably even suggest dark life coexisting with our own. In this scenario, rather than the usual animated creatures fighting other animated creatures or on rare occasions cooperating with them, armies of dark matter creatures could march across the screen and monopolize all the action.

But this wouldn't be too interesting to watch. The problem is that cinematographers would have trouble filming this dark life, which is of course invisible to us—and to them. Even if the dark creatures were there (and maybe they have been) we wouldn't know. You have no idea how cute dark matter life could be—and you almost certainly never will.

Though it's entertaining to speculate about the possibility of dark life, it's a lot harder to figure out a way to observe it—or even detect its existence in more indirect ways. It's challenging enough to find life made up of the same stuff we are, though extrasolar planet searches are under way and trying hard. But the evidence for dark life, should it exist, would be far more elusive even than the evidence for ordinary life in distant realms.

We have yet to directly see gravity waves emitted by a single object. Even black holes and neutron stars, which astronomers have detected in other fashions, have so far evaded gravity-wave detection. We stand little to no chance of detecting the gravitational effect of a dark creature, or even an army of dark creatures—no matter how close all of them might be.

Ideally, we would want somehow to communicate with this new sector—or have it correspond with us in some distinctive manner. But if this new life doesn't experience the same forces that we do, that's not going to happen. Even though we share gravity, the force exerted by a small object or life-form would almost certainly be too weak to detect. Only very big dark objects, like a disk extending throughout the Milky Way plane, could have visible consequences—like those discussed below.

Dark objects or dark life could be very close—but if the dark stuff's net mass isn't very big, we wouldn't have any way to know. Even with the most current technology, or any technology that we can currently imagine, only some very specialized possibilities might be testable. "Shadow life," exciting as that would be, won't necessarily have any visible consequences that we would notice, making it a tantalizing possibility but one immune to observations.

In fairness, dark life is a tall order. Science fiction writers may have no problem creating it, but the Universe has a lot more obstacles to overcome. Out of all possible chemistries, it's very unclear how many could sustain life, and even among those that could, we don't know the type of environments that would be necessary. Enticing as it is, dark life is not only hard to test for. It is hard for the Universe to create. I will therefore leave this possibility aside—at least for now—and focus on the targets and the searches for a big dense disk that I expect to be more promising.

SIGNS OF A DARK DISK

To be systematic and to start with the most minimal assumptions, JiJi Fan, Andrey Katz, Matt Reece, and I first investigated the simplest DDDM model we could think of. In addition to the usual feebly interacting dark matter, our model contained dark-charged particles and a dark force analogous to electromagnetism, through which the charged dark matter particles interact. The model included a heavy particle that is positively charged like a proton and another type of negatively charged particle akin to the electron.

Working on a novel idea that hasn't yet entered the physics canon is almost invariably a bit of an uphill battle. For some physicists and astronomers, double-disk dark matter is a stretch. Even for particle physicists, despite the rather daring nature of their research that aims to uncover the fundamental building blocks of matter, many colleagues—and scientists in general—are on the whole a conservative lot. This is not entirely unwarranted: if there is a conventional explanation for an observation, it is almost always the right one. Radical departures should be accepted only when they explain phenomena that older ideas fail to accommodate. In only very rare instances are new ideas truly necessary to explain observations.

Even when the scientific community agrees that something new is called for, deviating beyond the few "accepted" proposals for which the preponderance of work has lent weight can meet with resistance. Supersymmetry and WIMPs, for example, are often viewed by particle physicists as almost established, even though experimental evidence for them is as yet nonexistent. Only in the face of increasingly constraining data do many members of the community begin to acknowledge doubt and start to consider new possibilities for what lies beyond the established research canon.

Once a newer concept takes hold, everyone works it to death, figuring out and testing every corner of the space of parameters—even for hypotheses that are not yet proven to be true. But before an idea reaches that level, a lot of (often justifiable) criticism reigns. A few particle physicists—myself and my collaborators among them—simply try to keep an open mind in the face of uncertainty. We might favor some theories that we find more elegant or economical, but we don't decide what is correct—or what to work on—until data's arbitrating influence opens or closes a door.

My collaborators and I soon realized that interacting dark matter, which behaves very differently from non-interacting dark matter, should have distinctive observational implications. But given the initial motivation behind the DDDM proposal, I'll also briefly consider its implications for more conventional search methods, such as the indirect detection signal that first stimulated our research—as well as for one place where DDDM might resolve an issue facing conventional dark matter scenarios. I'll start by considering indirect signals, such as the Fermi photon signal that led to our research.

A thin dark disk is dense, meaning the concentration of dark matter particles is high. Within the dense disk, more dark matter encounters and hence more annihilations should occur than for dark matter distributed in a conventional cold dark matter halo. This doesn't mean that DDDM models will all be observable in this way. For DDDM to generate an indirect photon signal, which was the initial stimulus for our idea, an additional ingredient over and above the charged dark matter I just described would be required. Because a Fermi-like signal requires dark matter to turn into photons, which are a form of ordinary matter, an observable interaction will arise only if there is a particle that is charged under both the usual electromagnetic force and the dark one—the analogue of the person who both watches Fox News and listens to NPR or is signed in to both Facebook and Google+. If a particle charged under both types of electromagnetism exists, then dark matter could annihilate into photons by producing this intermediary particle that connects to both the dark and visible sector. This makes the Fermi signal a possible prediction, but not a generic one, of DDDM.

The dense disk does, however, mean that if observable interactions exist, they will occur at a faster-than-expected rate. The even better news is that if DDDM generates any indirect detection signal, be it photons or positrons or antiprotons, the result will be distinguishable from that of any other type of dark matter model. With the dark matter indirect detection signal from the usual type of dark matter, the prediction for the rate is highest near the center of the galaxy, where the dark matter density is greatest. The signal from DDDM would also be stronger toward the galactic center, but any signal that comes from the galactic center should exist throughout the entire plane too—since dark matter is dense throughout the region. Such visible annihilations throughout the galactic plane would be a smoking gun for DDDM.

Of interest too are the potential implications of DDDM for direct detection experiments, which are, after all, the holy grail of many dark matter seekers. Recall that direct detection relies on a small interaction between dark matter and ordinary matter, which allows for the deposition of tiny recoil energy that a detector can potentially record. As with indirect detection, any direct detection signal of DDDM models would also rely on the optimistic (and nongeneric) assumption that dark matter has some interaction with ordinary matter—feeble enough to be consistent with everything we know but strong enough to lead to a detectable signal.

The direct detection signal also depends on the local dark matter density, since, after all, the more dark matter the better. Disk dark matter may or may not exist in the vicinity of ordinary matter—it depends on the thickness of the dark disk plane—but if it does, it should have much greater density than dark matter in the halo.

It is also well known that the dark matter detection rate depends on the dark matter particle's mass, which helps determine whether the recoil energy is sufficiently big to be recorded and, if so, the amount of energy that would register. The detectability of the signal has a similar dependence on a more overlooked characteristic of dark matter, its velocity, which is also critical to the kinetic energy and hence the amount of recoil energy. Faster dark matter is easier to detect than slower dark matter since the energy deposited would be greater.

DDDM has much lower velocity in and out of the galactic plane than ordinary dark matter because it has cooled down. Moreover, the dark matter orbits around the galaxy in the same fashion as the Solar System, so its velocity relative to us is very small too. The low velocity of the new dark matter component relative to us means that DDDM would impart so little energy in a direct detection experiment—even if it did interact—that it would almost certainly be below the energy detection threshold and would therefore not be seen. Without more sensitive detectors or some additional ingredient in the model, conventional DDDM interactions would go unrecorded in the usual direct detectors.

However, experiments with lower thresholds are in the works and variants on the model might allow a signal even before their completion. What's interesting here too is that should a signal be seen, it would be distinctive enough to identify a DDDM origin. The low velocity of the dark matter would lead to a signal that was much more concentrated in energy than any other dark matter candidate that has previously been proposed.

One further interesting test of our model—or any dark matter model containing charged dark matter that combines into atoms—comes from detailed studies of the microwave background radiation. Several astronomers and physicists used the CMB and galaxy distribution data to look for evidence of dark atoms and DDDM in an interesting new way.

Remember, radiation in normal matter can wipe out density variations in charged matter—much as wind on a beach smoothes out the evidence of tides—while dark matter simply attracts further growth in structure. The distinctive influences that imprint themselves on the cosmic microwave background radiation can be used to distinguish dark and ordinary matter. Ordinary matter can also leave an imprint when charged matter combines into neutral matter, much as you see a distinctive rise in the sand at the maximum extent of the water creeping up onto a beach.

If dark matter—or at least some of it—also interacts with dark radiation, effects that are similar to those of ordinary matter will imprint themselves on the background radiation. Since our model contains both a heavy dark matter particle and a light one with opposite charge—very much akin to a proton and an electron—these particles would combine into dark atoms that would register in ways very similar to ordinary matter.

Detailed studies of the cosmic microwave background radiation showed that the fraction of dark matter that can have interactions of the sort we suggested is constrained. If the temperatures of the two sectors are reasonably similar, as would be the case if the dark and ordinary sectors interacted enough early on, the amount of interacting dark matter might have to be as low as five percent of the amount of total dark matter—about one-quarter the amount of visible matter. Fortunately this value is still interesting and it should also be observable using the method presented below. It is also well within the value needed to explain the periodic meteoroid strikes that I'll get to in the following chapter.

MEASURING THE SHAPE OF THE GALAXY

The research just described was interesting in that it demonstrated not only the power of the cosmic microwave background radiation, but also the significance of large data sets in the modern cosmological era—the processing of which astronomers are well set up to do. With the input of a model-building perspective and the technological and numerical advances currently under way, we have a much better chance of finding influences of unconventional dark matter—even when they are only subtle effects on the observed distribution of structure. My collaborators and I recognized that most interesting and robust signals are probably not the ones targeted by the usual dark matter searches that I've already described. More promising observational consequences of a dark disk follow from the gravitational pull of the disk itself. In today's era of "big data," the best places to look for distinctive dark matter properties could well be seemingly ordinary astronomical data sets.

The most obvious and decisive generic implication of the DDDM proposal is the existence of a thin dark disk in the central plane of the galaxy. If the dark matter particle is heavier than the proton, the disk will be narrower than the one containing stars and gas, making the gravitational potential exerted by the Milky Way galaxy—and all the others—different from what would be expected without the new form of dark matter. Like targeted advertising, the dark disk will add extra heft to the more diffuse ordinary matter component—and furthermore change matter's distribution—influencing the gravitational potential most dramatically near the galactic midplane, where the dark matter disk is concentrated. Because the gravitational influence of this matter distribution would influence the motion of stars, when enough positions and velocities of stars are measured with adequate precision, the distribution will confirm or rule out a dark disk (at least one with big enough density to make a difference).

One of the most incredible developments that JiJi, Andrey, Matt, and I learned about when we first started thinking about the dark matter disk in the summer of 2013 was that precisely this measurement in the Milky Way galaxy was set to occur. A satellite that was scheduled to launch that fall (or in the spring for those at the French Guiana launch site, as pointed out by my bemused Australian colleague) should measure this distinctive gravitational influence.

The GAIA satellite will in effect measure the shape of the galaxy. Within five years we will know the result. Preparations for the satellite were already well under way when we worked on our first paper, but it will conduct precisely the dark disk measurement we might have requested had we been asked during its preparation. In fact, although they didn't have our precise model or methodology in mind, astronomers had argued for the GAIA mission in large part based on its ability to determine the mass distribution in the galaxy—no matter what type of matter or where in the galaxy it is located. Though the takeoff was delayed a couple of months from the initially scheduled date, the launch in December of that year—only several months after we had finished our paper—certainly seemed remarkably fortuitous.

Particle physicists don't encounter this type of surprise very often. We know which experiments are possible and try to find if they can be tweaked or interpreted in a way that will test new ideas. Experimenters at the Large Hadron Collider (LHC) at CERN investigate some of the proposals that Raman Sundrum and I and others devised to explain the mass of the Higgs boson, for example. Though the LHC experiments were initially designed with other models in mind, Raman and I were fully aware of them and their potential when conducting our research on a warped extra dimension of space.

On the other hand, sometimes an idea is sufficiently compelling and testable that experimenters will respond and design a relatively small-scale experiment to rule out or verify the proposal, as when physicists designed experiments to measure the gravitational force more precisely in response to large extra-dimension ideas.

But rarely does it happen that an experiment just happens to be starting that is poised to test an idea that was studied independently for completely different purposes. Yet this is what materialized. The GAIA satellite houses a space observatory that will measure the positions and velocities of a billion stars of the Milky Way, with the goal of creating an extremely precise and extensive three-dimensional galactic survey. Its measurements will map onto a particular galactic potential and thereby tell us the galaxy's density distribution. If this distribution demonstrates the presence of a dark disk, the disk's thickness and density will tell us in turn about the mass of the new type of dark matter particle and how much of the interacting dark matter exists.

The method is based on an idea suggested by Jan Oort—the astronomer who also established the existence of the Oort cloud. Oort realized that the velocities of stars as they go in and out of the galactic plane depend on the shape and the density distribution of the disk, since their motion responds to the disk's gravitational pull. Measuring the velocities and the positions of stars oscillating in and out of the plane therefore pins down the density and spatial distribution of matter in the disk.

This is precisely what we would like to know to test or confirm our dark disk proposal. The gravitational attraction of a dark disk affects the motion of stars since they respond to the galaxy's gravitational pull. Knowing positions and velocities precisely for so many stars will reveal the galaxy's gravitational potential and establish whether or not a dark disk exists. With detailed information about the disk potential and the spatial distribution of matter within it, we can hope to determine more about the properties of the disk and about the interacting dark matter that created it.

But we don't have to wait for GAIA data to test the method and get a preliminary result. We already have useful data from the Hipparcos satellite, which the European Space Agency launched in 1989 and which continued to operate until 1993. Hipparcos was the first to do the detailed position and velocity measurements, but they did so with less accuracy and with fewer stars than GAIA will survey. Yet its results, though not as complete as GAIA's will be, already constrain the form that a dark disk might take.

This insight, though new to us as particle physicists, was well known among some astrophysicists. In fact, using this method, a few researchers had even gone so far as to conclude that existing data already rules out a dark disk. This cavalier disk denial confused many people, including one of our paper's referees. A moment's reflection, however, tells you that this result (at least as stated) is not possible. No matter how accurate the measurement, the density can always be sufficiently low to evade any existing bound. What astrophysicists were really saying was that there was no need for a dark disk. Given the uncertainties in densities in all the known gas and star components, the measured potential could be accounted for by known matter alone.

But sometimes the right question is what else might be consistent, and therefore be a viable alternative interpretation of the data. The only way to find out whether something is allowed or even preferred is to evaluate the consequences of new assumptions and determine their experimental implications. My collaborators and I asked a different question from the astronomers. We didn't ask for proof that a dark disk exists. The real question is how substantial a disk can be while maintaining consistency with all observations. And whether perhaps the introduction of a dark disk component might even match the data better.

This different way of thinking reflects in large part the difference between the sociology of particle physicists—particularly model builders—and many astrophysicists. To give credit where it's due, the astrophysicists taught us quite a lot. We learned how they approach the problem and what data currently exist. Their methods are extremely useful. But approaching a problem from a different angle often leads to new insights and opens up new possibilities. Whether or not a dark disk exists, we will only know by making the assumption that it does and figuring out what is allowed. Everyone wins in the end.

We wanted to know if a dark disk is allowed, or even possibly favored, by the data—not just whether or not one can fit the measured stars' properties without one. Each of the ordinary matter components that is added to compute the Milky Way disk's gravitational potential is known only so well. Allowing for the uncertainties in the measurements certainly creates room for something new. This is the task I set out for a student, Eric Kramer, who studied the Hipparcos data as well as gas density measurements in the galactic plane. Together we identified many assumptions that went into the astrophysicists' analysis that needed to be revisited. Although a cursory examination of the Hipparcos results could prematurely lead to the conclusion that a dark disk was disfavored, a more careful and up-to-date analysis demonstrated that the data didn't suffice to make such a claim.

The Hipparcos data itself provide some of the uncertainties. But the relatively poor measurement of some of the visible matter in the Milky Way is a major source of uncertainty too. The more wiggle room there is, the more room there is for a dark disk. On top of that, because all components of matter experience the gravity exerted on them by the other components, only by including all matter—including the dark disk—from the beginning can one extract the true constraints. This is one of the merits of having a model. It gives a well-defined target and a fixed computational strategy when evaluating the results of a search.

With a careful analysis, we found that there is room for a dark disk. Indications are promising, but before more conclusive data becomes available we won't know if DDDM models will prove to be correct or whether simpler, more standard scenarios suffice to account for the matter in our Universe.

This brings me to the question: what is the dark disk density we hoped to target in the first place? That is, how strong a constraint would be interesting? From many perspectives, any value is worth pursuing. Finding a dark disk, no matter how low its density, would be a fundamental change in our view of the Universe. But we will soon see that another target comes from the periodic-meteoroid-strike-inducing aspect of the dark disk. I will simply say for now that the value we found that was necessary to trigger meteoroid attacks is consistent with current data.

Furthermore, although it wasn't our original intention, partially interacting dark matter might also help solve some outstanding mysteries of more conventional cold dark matter scenarios. The astronomer Matthew Walker, now a professor at Carnegie Mellon University, suggested that DDDM might help address the problem with dwarf satellite galaxies in Andromeda alluded to in Chapter 18. A world with ordinary matter or even conventional cold dark matter offers no explanation should these results hold up. A Harvard postdoctoral fellow, Jakub Scholtz, and I showed that self-interactions for a component of the dark matter might be the unique solution to the problem of how dark-matter-dominated dwarf galaxies that are aligned in a plane are formed. Jakub, Matthew Reece, and I are also investigating the potential implication of DDDM for primordial black holes, which are bigger than they should be in standard scenarios.

The Fermi gamma ray signal that motivated our project now appears to have been a red herring since the signal has faded over time. But the dark disk scenario that grew out of trying to understand it has wide-ranging implications that should make DDDM observable in other ways. The scenario might even have more interesting implications for galaxy formation and dynamics that we can now begin to explore.

So, after our expansive exploration of the cosmos and the Solar System, let's culminate our journey by bringing together a lot of these ideas. We'll now consider how dark matter might affect us close to home—affecting the motion of stars and potentially the stability of objects at the outskirts of the Solar System.

21

DARK MATTER AND COMET STRIKES

"Boffins" is not a term that is familiar to most Americans. So when the science and technology writer Simon Sharwood conferred this distinction on me and my collaborator Matthew Reece in the British journal the *Register*, I wasn't initially sure what to think. Was the author criticizing us and our foolish ways, or was "boffins" a term like "pulchritude," which sounds pretty bad—but is actually a term of flattery?

I was comforted to learn that "boffins" just means scientists or technical experts—albeit possibly overly focused ones. But my initial fear that the word might have meant "loony people" or some such was not entirely unfounded given that the subject Sharwood reported on was our work about dark matter and meteoroids—with a brief shout-out to extinctions. The idea was that dark matter could effectively sling comets out of the Oort cloud so that they periodically catapulted into Earth, possibly even precipitating a mass extinction.

Even for particle physicists like Matthew and me who try to keep broad and open minds, messy phenomena like meteoroid strikes tied to the complicated dynamics of the Solar System as a whole and to dark matter on top of that seemed like an uncertain route. On the other hand, dark matter (!), meteoroids (!), dinosaurs (!). The five-year-olds in us were intrigued. As were the adults who were curious to learn more about the Solar System. Not to mention the scientists we are, who wanted to know if we could actually learn something more about these various pieces and how they might all tie together. After all, although we have yet to measure the presence of a dark disk, a satellite that measures a billion stars in the Solar System could have the sensitivity to decide the issue within the next five years—and test whether our proposal is correct.

Just in case this scenario, the richness of the ideas, or the upcoming satellite measurements weren't argument enough to pursue this line of inquiry, the day I asked Matt if he wanted to consider the project was the day that the Chelyabinsk meteoroid struck. Although most of the many meteoroids hitting the Earth or its atmosphere are so small that we don't usually notice them, the particular meteoroid that exploded on February 15, 2013, was 15 to 20 meters wide—large enough to glow brilliantly and release the equivalent of 500 kilotons of TNT. That this meteoroid exploded three days after a question from the University of Arizona audience had started me thinking about periodic meteoroid impacts—and the day on which I had proposed to Matt that we delve deeper into the subject—struck us as remarkable, and pretty funny. We were wondering if we should study why extraterrestrial objects reach the Earth and on that very day something did. How could we not proceed?

I will now describe our research that ties together many of the ideas this book has explored and explain how dark matter could affect the planet on an approximately 30 to 35 million year time scale. If we are correct, not only did a roughly 15-kilometer-wide meteoroid hit the Earth 66 million years ago, but the trigger for that impact was the gravitational influence of a dark matter disk in the midplane of the Milky Way.

THE SCENARIO

We now have a picture of the Milky Way galaxy with its bright disk of gas and stars, and inside, perhaps another, denser disk composed of interacting dark matter. The Milky Way came into existence more than 13 billion years ago when dark matter and ordinary matter collapsed into a gravitationally bound structure. Perhaps a billion years after the galaxy halo formed, ordinary matter radiated away energy to begin to form the brightly lit disk we now see. If some of the dark matter interacts and radiated dark photons sufficiently quickly, it too collapsed into a thin planar region we call a disk. This might have all taken a while to complete, but the narrow dark disk would have been established long ago.

Meanwhile, roughly four and a half billion years in the past, the Sun and the Solar System were formed. Planets subsequently emerged from the disk of matter that circled around the Sun. After the planets' formation, Jupiter moved inward and other giant planets moved outward and, in doing so, scattered material in the disk. Some of that material moved very far away to the distant region of the Oort cloud, where small icy objects are bound to the Sun only by a very weak gravitational pull.

The Solar System then circled around the galaxy every 240 million years. But with a period of perhaps 32 million years, the Solar System, while traveling along its dominantly circular path, also bobbed up and down through the galactic plane. The gravitational pull of the disk acted on the Sun throughout this journey, serving as a restoring force every time the Solar System escaped as far as it could in the vertical direction above or below the plane. Because the galaxy provides so little friction, the Solar System repeated its vertical motion through the galactic plane on a periodic basis, with the restoring force of the plane acting on it every time that it passed through.

Moreover, when the Solar System was in or close to the galactic plane, the distorting gravitational tidal effects of the disk acted on it most strongly. During these particularly strenuous intervals, the tidal influence of a thin dense disk of dark matter might have disrupted the tranquility of some of the weakly bound objects in the Oort cloud, which would otherwise have continued mostly undisturbed along their distant orbits. Once in range of the dark disk, Oort cloud's icy objects were unlikely to all stay in place in the face of this bumpy ride.

In the meantime, while all these inanimate objects were going about their business, life on Earth began to form about three and a half billion years ago and complex life proliferated about three billion years later—540 million years in the past. Life has had its ups and downs since then, as diversification competed with extinctions. Five major extinctions punctuated this time frame known as the Phanerozoic era. The last of these occurred 66 million years ago, when a meteoroid strike devastated the Earth.

Up to the time right before impact, the dinosaurs were oblivious to any distant Solar System snafus. Icy bodies orbited through the Oort cloud, with only the occasional small change to their orbits from the distant tug of the Milky Way disk, which acted with varying strength according to the Sun's distance from the midplane. The orbits of a few of these icy bodies became so distorted that their paths took them to the inner Solar System, where some broke away from their initial trajectories under gravity's distorting effect. At least one of these icy bodies might have turned into a comet on a collision course with the Earth.

From the perspective of the Oort cloud, it was a relatively minor disturbance. One or at most a few of its icy bodies got dislodged. But from the perspective of 75 percent of life on Earth, including the venerable dinosaur, the meteoroid that was set to strike was apocalyptic. Yet even if the dinosaurs had been sentient, conscious beings, they wouldn't have noticed anything extraordinary was about to happen when the comet first appeared. Though the comet's nucleus was bright enough to see during the day and its long tail was visible throughout the night, the comet would have betrayed no notable sign of the imminent devastation it was poised to deliver. This impression might have changed as the comet descended, when fire and debris lit up the sky. But whatever the doomed creatures might have seen or envisioned, once gravitational influences had altered the comet's course, the fate of those animals was irrevocably sealed.

The comet would shortly slam into the Yucatán, pulverize its target, and end a journey that would culminate in massive global destruction. When the meteoroid whose impact created the Chicxulub crater struck, the impact vaporized the comet as well as the ground near where it hit—sending up plumes of dust that scattered all around the globe. Fires burned the Earth's surface, tsunamis flooded coastlines both near the impact and on the opposite side of the planet, and poisonous materials rained down even more dangers. The food supply was decimated, so any land-dwelling creatures that did manage to survive the immediate aftermath probably starved to death in the weeks and months afterward. Most life didn't stand a chance when confronted with such sudden and drastic changes to the Earth's climate and its various habitats. Only ground-burrowing mammals and airborne birds remained when conditions eventually improved to perpetuate advanced life into the uncertain future of a very different age.

It's a dramatic picture, and the basic facts of the meteoroid impact are by now well established. The many observations of geologists and paleontologists have confirmed that a big object hit 66 million years ago and that at least 75 percent of life on the Earth then died. Shortly I will describe how a dark disk might have been the trigger responsible for dislodging the comet that was responsible for all this devastation. But first I'll explain the genesis of the idea.

INCEPTION

Many benefits accrue from sharing physics with the general public through books and lectures. But because time invested in these activities can detract from ongoing research, I frequently have to prioritize and choose which talk requests I should accept. However, on some happy occasions, my research profits from what I had wrongly feared would be a distraction by taking me to visit people I would not have ordinarily encountered or by introducing me to an idea I might otherwise not have considered.

In February 2013, I reaped such a reward from the astrophysicist Paul Davies's invitation to give the Annual Lecture he hosts at Arizona State University's BEYOND Center. Despite my hesitations about too much travel, ASU has a very strong cosmology research group so I was happy to agree not only to give the public lecture, but also to present a seminar the following day to the experts in the department there. The seminar would be more focused on my recent research, which was the double-disk dark matter idea I have already described.

The physicists in attendance asked several excellent questions about the model—its detectability and its implications for the cosmic microwave background radiation, for example. But I was thrown for a loop when Paul asked me if the dark matter disk was responsible for the dinosaurs' demise. I confess that at the time, I hadn't thought much—really ever—about dinosaurs in my scientific research, which had been focused on elementary particles and elements of the cosmos. But Paul informed me of potential evidence for periodic meteoroid strikes and the absence of a good explanation for them. He asked whether a dark matter disk might fit the bill—and in the process reminded me about the meteoroid that had extinguished the terrestrial dinosaur.

Paul's question was too good to ignore. The answer wasn't straightforward and I had a lot to learn before more definitively responding, but dark matter and the dinosaurs certainly seemed like a connection that could teach me—and potentially scientists more broadly—quite a lot. I asked Matthew Reece whether he would be interested in studying the possibility that meteoroid strikes were triggered by our proposed dark disk, which sounded more connected to physics than a question about the dinosaurs.

Matt was the obvious choice of collaborator. He played a crucial role in the initial DDDM research, has a cool technical head, and is scientifically open-minded when it comes to new ideas—more so than you would anticipate from his decidedly conservative demeanor. He doesn't make the common fallacy of assuming anyone—even overly confident more senior colleagues—has correctly guessed everything.

But most important, Matt is an excellent physicist with very high scientific standards. When he does something, you can be sure it is on solid footing. Still, I wasn't sure how he would react to such an apparently crackpot suggestion. I was very pleased when Matt found the idea intriguing and recognized its potential scientific merit. Paul Davies was interested too, but he already had many ongoing research projects and graciously chose to be in contact but not to participate.

So after listening with amazement to news about Chelyabinsk on the very day we started discussing the idea, Matt and I pressed on to see what we could learn. Our goal was to turn this crazy notion of a dark disk causing meteoroid strikes into testable science. As model builders and particle physicists, Matt and I try to be receptive to new ideas and interpretations. But we also are fully aware of the importance of remaining unbiased and careful. These qualities were essential in the research I'll now describe.

THE DARK DISK AND THE SOLAR SYSTEM

As explained in Chapter 14, in order to be realistic in our goals, Matt and I decided first to pare down our investigation. Despite our curiosity about dinosaurs, we initially left aside the additional challenges endemic to extinctions and focused solely on meteoroids and Solar System dynamics and a possible periodicity in the physical crater record. With the extinction issue on the back burner, we could directly investigate the dark disk's potential influence on comets and whether it might be responsible for periodic meteoroid strikes. We could decide afterward how well our meteoroid predictions explained any particular known impact, including the one responsible for the K-Pg extinction.

We then made sure that none of the previous proposals for periodic triggers that might dislodge Oort cloud objects could successfully explain a periodic signal. If a more conventional mechanism sufficed, then no one, ourselves included, would bother evaluating the consequences for the crater record of a more exotic scenario—no matter how cool and seductive it might be.

However, as explained in Chapter 15, conventional triggers don't work. With only the standard Milky Way disk, the tidal effects from the galaxy are too smooth and perturbations from stars are too infrequent. Neither the usual tidal effects, Nemesis, Planet X, nor the Milky Way's spiral arms suffice to trigger frequent or notable enough comet showers. These earlier proposals didn't yield either the correct time between galactic plane crossings or sufficiently sudden strikes to match the crater record. For example, with only normal matter in the disk to influence the motion, the Sun's vertical oscillation period would be more like 50 or 60 million years—too long to match the available data.

This left two possible conclusions: either the periodicity wasn't real, as might well turn out to be the case, or the more interesting logical alternative is correct and the trigger is unconventional. With the previous suggestions ruled out, it made sense for us to ask whether our proposed dark disk could succeed where ordinary matter alone failed and give rise to the required periodicity and change in rate. Indeed, the dark disk has just the necessary properties to address the inadequacies of the normal matter disk. With a thin disk of dense dark matter, the disk tidal force can successfully account for both the period and time dependence of perturbations to the Oort cloud.

Recall that throughout their existence, Oort cloud objects are subject to the disk tidal force from ordinary matter and occasionally to the more intermittent but nonetheless important influence of passing stars. These effects serve to move around the distant, relatively weakly gravitationally bound bodies of the Oort cloud and nudge them closer to the Sun. The tidal effect of the Milky Way plane can then give one final tug that might bring these icy bodies into precarious, very eccentric orbits that jut inward to about 10 times the Earth-Sun distance, where the gravitational pull of the large planets will very likely remove them from the Oort cloud. This pull will either fling such comets out of the Solar System or pull them in so that they enter tightly bound orbits in the inner Solar System. These disruptions account for the generation of long-period comets, with several new ones entering the Solar System every year. Occasionally, however, perturbed objects get deflected out of their orbits altogether, and at those times deviant comets might strike.

But this type of perturbation does not in itself suffice to explain periodic impacts. For periodic strikes to occur, a rapid change in the rate of disruptions to the Oort cloud must occur at regular intervals. Furthermore, to match the available evidence, the period must be in the range of 30 to 35 million years. If even one of these criteria fails, a proposed explanation for periodic meteoroid strikes won't do. And neither criterion is satisfied for any conventional suggestion.

However, the addition of the denser and narrower dark disk addresses both these issues exceedingly well. Once you accept the possible reality of periodic meteoroid strikes, a dark disk is indeed a very promising idea. The dark disk exerts an influence that is both more intense and more rapidly varying with time than the conventional disk of the galactic plane—the two essential requirements to create spikes in comet intensity.

With the dark disk included in the Milky Way plane, the Sun's vertical oscillation period would be shorter than the period induced by the conventional Milky Way disk alone because the gravitational pull with the addition of the dark disk's matter is stronger. On top of that, according to current matter density determinations, the Solar System oscillates only about seventy parsecs above and below the galactic plane—a much more limited range than the thickness of the full ordinary matter disk. The narrow dark disk, which therefore would encompass the Solar System throughout a reasonably large fraction of its trajectory, can have a disproportionately large influence on the Solar System's motion as it bobs up and down through the plane.

The other merit to a thin dark disk is that, even so, the Solar System can pass through it quickly enough to induce a spike in the comet rate that lasts a million or so years. Because of its big time-dependent influence, the dark disk triggers further perturbations each time the Solar System crosses the galactic plane, creating comet showers on a regular basis—during each plane crossing—that otherwise would be triggered only very infrequently by closely approaching stars. The enhanced tidal effect takes place when the Solar System crosses the narrow region occupied by the dark disk. Only during this passage and perhaps for one or two million years following will comet strikes be enhanced.

When the Solar System passes through the disk on this time scale and is subject to an enhanced tidal force—a spike in the force if it happens sufficiently quickly—icy bodies in the Oort cloud might get dislodged and a few might even come hurtling into our planet at about 50 km/sec. Once set off track in this manner, the trip is quick—perhaps a few thousand years. But the perturbation that set it off in the first place takes place more slowly—generally taking one to a few orbital periods to get going. This means that in a time period between about one hundred thousand and a million years, the fate of comets that have come too close to the Sun will be determined, and some of those could account for comet showers that hit the Earth's atmosphere or the Earth itself.

Matt and I worked out the predicted trajectory and the scenario was a success—at least within the confines imposed by the limited and somewhat shaky data. But there was one final check that we hadn't completed, as was pointed out to us by the referee who reviewed our paper for the prestigious physics journal *Physical Review Letters*. In addition to determining the motion of the Solar System in the presence of the dark disk, we calculated the rise and fall in density of the Solar System's environment as it passed through. We needed to know the density since we assumed that whatever disturbed the Oort cloud would be proportional to this degree of matter concentration. After all, more mass means more tidal influence, which should mean more disturbances. We therefore assumed that density would serve as a useful proxy for the rate of meteoroid strikes, as indeed turned out to be the case.

But we hadn't yet explicitly confirmed that the tidal distortion of the Oort cloud by which the dark disk acts on the icy bodies in the cloud was sufficient to rain down comets at the correct rate. Fortunately for us, Scott Tremaine and Julia Heisler had already done a lot of the heavy lifting about a decade earlier, so we could simply apply their results. And indeed our assumption had been right. The enhanced density creates just the sort of pull needed to dislodge comets on the appropriate time scale.

I was actually impressed by the referee's useful suggestion. These days, referee reports, in which colleagues who should be experts review papers before approving them for publication, are often either rubber stamps or vehicles for aggrieved authors looking for citations. This referee's suggestion actually taught us some physics. The tone had been dismissive, but we learned something from following it up. We had to deal with some misguided criticisms too—but since we had been careful to check papers and experts beforehand, we could readily pinpoint the flaws in those critiques.

In the end, Matt and I calculated the preferred density and thickness values for a dark disk that matched the crater record, and found that they were in line with our previously existing DDDM model, which by that time we knew was consistent with existing galactic measurements. The new, even better wrinkle that Matt and I found in our research was that not only was the dark disk allowed, but it was actually favored if you take the disk seriously as the instigator of comet strikes.

The dark disk's surface density should be about one-sixth of that of matter in the ordinary disk. That's enough to be interesting but not so much as to overwhelm any currently understood phenomena. It's a sizable chunk of the dark matter—not one-millionth, for example, but instead a few percent. Should this dark component exist, it would be sizable enough to have measurable influences and therefore would be worthy of attention. Furthermore the disk thickness might be less than one-tenth the thickness of the ordinary matter disk—less than a few hundred light-years in width compared to ordinary matter with a roughly 2,000 light-year thickness. Again, it is this narrowness of the dark disk that explains why it can conceivably trigger dramatic effects on a periodic basis.

We found the dark disk with the right density was favored by a factor of three. A key contributor to this new conclusion with better statistical support was the look-elsewhere effect that I mentioned in Chapter 15. With a definite model for what might trigger periodic strikes, we could not only better predict the period, but we could also do so more reliably. In fact, our intention in the paper went beyond demonstrating that a dark disk could succeed in explaining periodic comet showers in a way that ordinary galactic components can't. We wanted to make a second point, which had to do with statistics, and how to evaluate the significance of these or any other results.

As discussed in Chapter 14, most existing searches for periodicity tried to match a periodic function for the up-and-down motion of the Solar System—a sine wave, for example—to the data. This can be interesting, but it doesn't capture the full story. We don't have to guess the motion of the Solar System. If we knew everything about the galaxy and the Sun's initial position, velocity, and acceleration, we could use Newton's laws of gravity to compute the Sun's motion and predict the period we should expect. After all, the motion of the Solar System isn't random, but has to be consistent with its underlying dynamics. Even with imperfect knowledge of the density distribution and the parameters of the Sun, the range of possible trajectories—and hence possible periods—is restricted.

Matt and I incorporated what we know about the densities of known matter in the galactic disk—allowing for the full range of possible values supported by current measurements—and added the matter contribution of a dark disk. Our goal was to check whether there was evidence for periodic cratering rates that matched the Solar System's motion once we took into account everything we know about the measured disk components—stars, gas, etc.—plus the dark disk component we had introduced.

The measured contributions of ordinary matter restrict the range of possible trajectories that the Solar System can take since the gravity of the matter in the disks—both ordinary and dark—acts on the Sun and influences its motion, thereby reducing the influence of the look-elsewhere effect. Matt and I used the measured densities to predict the periodic motion of the Solar System and compared the times of galactic plane crossings to the reported crater creation times to check how well they match. Although predictions without an underlying model don't discriminate sufficiently, we found that with existing measurements accounted for, the statistics do favor periodic meteoroid onslaughts with a period of about 35 million years. Recent data improvements indicate the period is likely to be a bit shorter even—perhaps 32 million years.

The dark disk was critical to making the scenario work and generating the favored impact rate. Turning the story around, with the better match of crater data to Solar system motion, a dark disk is actually preferred. Future data should be analyzed with this sort of model in mind in order to yield the best statistical significance. The results will then either strengthen our result—or rule it out.

AND THE DINOSAURS . . .

After Matt and I had sorted everything out and our research had been accepted by *Physical Review Letters,* we posted our results to the online journal repository that provides immediate Internet access to the research papers known as preprints that are yet to be published. Matt did the actual submission. We had conservatively titled our paper "Dark Matter as a Trigger for Periodic Comet Impacts." But to my surprise, Matt had edited the comments section—generally used to describe format or revisions to the submission—to read "4 figures, no dinosaurs." I thought this was pretty funny since we had studiously avoided any explicit mention of dinosaurs in our paper, which focused on the crater record and its more direct contact to physics. But of course we had had this connection in mind all along and even jokingly referred to our work as "the dinosaur paper." I suppose that had I paid more attention, I wouldn't have been surprised the next day by the degree of online interest in our work, which was reported in many blogs and journal websites—including the "boffins" piece—almost always accompanied by some rather entertaining graphics.

But this does bring me back to the dinosaurs. Having established at least a first attempt to put data together with models to predict comet impacts, and knowing that this is not the final word but will be improved with future measurements, we did then look to see how well our model agreed with the timing of the Chicxulub event. Our calculations showed that, depending on the improving measurements of ordinary matter in the Milky Way disk, meteoroid strikes should occur about every 30 to 35 million years. Since we passed through the galactic plane within the last two million years, a comet dislodged from the Oort cloud one complete oscillation (two disk crossings) in the past might indeed have come hurtling down to Earth 66 million years ago, at the time of the K-Pg extinction, to wreak its enormous destruction. As an aside, if we passed within the disk less than a million years ago, we could even be on the tail end of an enhanced comet flux and have the potential to see heightened impacts today. But much more likely is that, aside from a truly random and extremely unlikely event, the Earth passed through a little further back and we won't witness another Chicxulub for about another thirty million years.

Because of the uncertainty in the Sun's position and the lack of knowledge of the precise period, we can only approximately predict the disk crossing times. If the Earth crossed the galactic midplane about two million years ago, an oscillation period of about 32 million years would be optimal for generating an event that occurred 66 million years in the past. Our original crude analysis produced a period of 35 million years, which is a little too big to match the Chicxulub timing—though the uncertainties in the model and in the length of time of enhanced comet strikes still permitted reasonable agreement. Our updated model for the Milky Way disk, which takes into account the more recent measurements of the galaxy's components, has since brought the period down—leading to a better match to the K-Pg extinction time. But even with the crude model we used for our initial prediction, there is a reasonable probability that the dark disk prediction corresponded to the Chicxulub event.

The primary reason that our results were not yet sufficiently precise is that the matter measurements in the Milky Way changed since we did our initial analysis. We also still haven't fully modeled the time-dependent galactic environment, such as galactic arms, which are also only poorly known. The density variation from these effects would not suffice to trigger meteoroid strikes. But they might well be sufficient to change by a few million years the precise prediction of the model for when those strikes would occur.

Other factors too contribute to the uncertainty in the exact times predicted for enhanced comet showers. It takes about a million years for the Solar System to cross the galactic plane—longer if the disk is thicker. Furthermore, a time period of up to a few million years might separate the initial triggering event from the actual meteoroid hit on Earth. Thirdly, the crater record and the dating precision are poor. Finding more craters or dating them more precisely would help—though new crater discoveries emerge only infrequently. Not just craters, but dust that gets trapped in rocks, might also help generate a more precise record of when comets have struck.

Evidence for 30 to 35 million year periodicity in the vertical motion of the Sun away and toward the galactic plane might come from unexpected directions too. After Matt and I had written our paper, a particle physics colleague who was aware of my fascination with astronomy, geology, and climate—but who, at the time, didn't know about the "dinosaur paper"—fortuitously told me about the work of Nir Shaviv, from the Hebrew University of Jerusalem, and his collaborators, who studied climate over the entire 540-million-year Phanerozoic era. Remarkably, they had found a variation of climate with a period of 32 million years—strikingly similar to the period that we had identified. If Shaviv's result holds up and this periodicity in climate is indeed determined by the Sun's motion through the galactic plane, the 32-million-year period too would be evidence of a dark disk since ordinary matter alone wouldn't suffice to yield this relative short interval between disk crossings.

Of course, we don't need to delve into the past to see the influence of dark matter. If dark matter does indeed have an interacting component that changes the structure of the matter distribution in the Universe, we will learn about that soon—perhaps even before any of the other dark matter searches reach fruition. Only a limited range of dark disk densities can account for the crater data. Future measurements will almost certainly narrow the range of possible predictions, validating or ruling out our proposal.

The analysis my student Eric and I have already done shows that the dark disk with the necessary density and thickness is permitted by observations to date. And the better data to come from GAIA will pin down a disk's presence, density, and thickness still further.

Once this satellite completes its 3-D map of stars in the nearest region of the Milky Way, the dark disk—or absence thereof—will be much better determined. By this indirect route, we might learn much more—not only about the galaxy and dark matter, but about the Solar System's past as well. If GAIA data establish the existence of a disk with the right thickness and density, it will be powerful evidence of the crater proposal's viability.

A better punch line would of course be that we precisely calculated the exact date of the dinosaurs' demise with sufficiently small uncertainty to be confident in the result. But this is a complicated subject involving many challenging measurements. Even so, the amount of progress scientists have made in the last 50 years is nothing short of astounding. Dark matter has been more elusive in many respects than the more readily apparent Earth, the Solar System, and the many other visible elements of the Universe. But through the research I've described, physicists are finding new ways to track it down. Whatever the outcome, we can be sure that the galaxy and the Universe, and the inner workings of matter itself, have some fascinating surprises in store.

CONCLUSION: LOOKING UP

I have had the good fortune to be invited to conferences with leaders in diverse fields—ranging from business, law, and foreign policy, to the arts, media, and of course science. Even when I have a different perspective from that of the other panelists or speakers, the discussions invariably stimulate fresh thinking about a wide spectrum of important topics. But the best questions—especially about my research—don't always come from the conference participants. A particularly gratifying exchange about physics at one recent event took place right after the conference had ended, when Jake—the young driver who was taking me to the local Montana airport—surprised me with his thoughtful interest.

Upon learning that I'm a physicist, many people feel compelled to tell me their attitude toward the subject—whether it's love or hate or fascination or confusion. I find this a bit funny. After all, most of us don't feel the need to inform lawyers, for example, about our thoughts on jurisprudence. But these curiously confessional physics conversations sometimes pay off. Jake explained to me how a few years back, he had relished college-level physics in his Oregon high school and how he was now eager to keep learning more. Although he was no longer taking classes, he wanted to hear more about the many advances that physicists have made since then toward our understanding of the Universe.

But Jake didn't ask only about recent developments. He also wanted to know about the place of the physics he had studied in light of later progress. So I explained to him that advances of the twentieth century taught us, for example, that Newton's Laws—though they remain an extremely accurate approximation in familiar environments—cease to apply when applied to high velocity, small distance, or high density environments in which special relativity, quantum mechanics, or general relativity take over.

After pondering this for a while, Jake posed an off-beat, but profound question. He asked me what I would do with my knowledge if I could go back in time, wondering if I would tell the people I met there about the more recent developments that we know about only now.

Jake recognized the two important aspects of this dilemma. The first was whether anyone would believe me—or if they would just assume I was nuts. After all, without the supporting experimental evidence that was obtained only through much more advanced technology, the remarkable phenomena and connections that scientists discovered and deduced in the last century might seem crazy. They violate the intuitions formulated in more commonplace surroundings.

But the second facet of his quandary was perhaps even more compelling. Jake wondered whether, even if people did listen and believe the new insights, they would be scared and ignore them or—at the other extreme—rush to apply them too hastily. His first instinct was to think I should keep the information to my imagined time-traveling self, reasoning that the world would be better off if it were allowed to evolve the way it had—with no shortcuts to scientific knowledge.

Given society's usual resistance to long-term thinking, Jake was worried that a sudden burst of information might be dangerous. He didn't think that change was bad. But he was apprehensive when he saw his younger siblings trapped by video games and smartphones—forgoing exercise and the outdoors and the sense of exploration which he had so enjoyed at their age. He also worried about the example of his home town, where he had seen industries rush to grab resources once new technologies were introduced, without regard for either the local or the global implications. Once Jake had reflected on the irreversible consequence—to the landscape and to his family's way of life—that he had seen take place during his short lifetime, he reasoned that society probably would be better off with sufficient time to adjust to major scientific discoveries or technological changes by more thoughtfully developing comprehensive, long-term strategies.

This book has explored how several major, uncontrollable perturbations in the Earth's past have profoundly disturbed the stability of our planet. One such disturbance with an extraterrestrial origin occurred 66 million years ago when a speeding comet—which might have been triggered by dark matter—precipitated a major extinction. Perhaps in another 30 million years or so, another one will do so again. Such events are fascinating to decipher, which is why I have focused on them in this book and continue to study them in my current research.

But understanding their impact on the planet and on life might have further benefits—helping us anticipate the consequences of some of the perturbations that we are making to our environment today. On the time scales relevant to civilization and the diversity of current life on Earth, a deviant comet is not our most pressing concern. But the changes that an exploding human population makes when it too hastily exploits the Earth's resources might well be. The impact might even be compared to that of a slow-moving comet—but this time the impact is of our own making. In contrast to an impact triggered at the farthest reaches of the Solar System—with the changes that are occurring now, we have the potential to exercise some control.

The study of dark matter is hardly the most obvious avenue to such concerns. *Dark Matter and the Dinosaurs* is a book about our surroundings in the largest sense—our cosmic environment and the remarkable insights that scientific advances have already brought us—and what future advances may come to reveal. But thinking about dark matter led me to think about our galaxy, which led me to learn more about the Solar System, which led me to consider comets, which led me to a better understanding of the extinction of the dinosaurs, which led me to contemplate the delicacy of the balance that allows life—meaning the life that exists on Earth today—to thrive. If we mess with that balance, we might survive and the planet certainly will. But it's less clear that the species we live with and rely on will survive any consequent radical changes.

The Universe has been around for 13.8 billion years and the Earth has taken some four and a half billion orbits around the Sun. Humans have graced the planet for a mere two million years and civilization for less than twenty thousand. Yet in my lifetime, the human population has more than doubled, adding more than four billion people to the planet. When we are too hasty in exploiting the Earth's resources—significantly influencing the planet and its life—we are very rapidly undoing the cosmic work of millions or even billions of years. The threats might escape ready notice in the short span of a human life. But exercising caution when moving forward could help us figure out more optimal ways to utilize new information and advances in the future.

We like to think we are resilient, but most likely the current state of the world is less stable than we think. Altering and destroying habitats and the atmosphere at the current rate is affecting biodiversity, and might even be precipitating a sixth extinction. Although humans certainly won't disappear any time soon, important aspects of our way of life might. The modifications we are making—or even our attempted solutions—threaten our environment, not to mention our social and economic stability. Though the consequences might ultimately be beneficial in some global sense, they won't necessarily be so for the species on Earth who live here now.

We can try to engineer some aspects of our environment but the world is an enormously complicated system that employs many miraculous-seeming features—only some of which we currently understand. Even if technology can solve some problems, it will be difficult in any case for it to keep up with the ever-increasing rate of change. Without repeated innovations to substantially alter the equation, the result will inevitably be an unsustainable expansion in which something has to give. A supportive political, social, and economic climate in which technology can be integrated into a more comprehensive strategy is essential if we are to achieve the optimal response. The challenges are clearly daunting but shouldn't prevent our making some headway towards this very worthwhile goal.

Exponential growth starts relatively slowly but then shoots up dramatically. The resources required to sustain this new status quo swamps anything we've encountered in the past. In our delicately balanced ecosystem and with our complex and fragile infrastructure, even relatively small disturbances have the capacity to generate large effects. It's important to ask ourselves whether we should plan our growth differently or at least anticipate the conceivable changes in a more deliberate manner. Even Pope Francis in his 2015 Encyclical warns about the faster and more intense human activity of what he calls "rapidification." Though some aspects of coming changes will be beneficial, the potentially detrimental consequences are worth anticipating too. Viewed from outside—or inside—we can seem pretty shortsighted.

Don't get me wrong. I believe in progress. After all, knowledge is a wonderful thing. But I also believe in taking responsibility for applying advances wisely, which sometimes means taking the long term view. An intelligent species shouldn't predicate its existence on competing for and destroying scarce resources that have taken billions or at least millions of years to emerge. Although technological applications can be useful or harmful—sometimes inadvertently so—increased knowledge gives us the ability to create desirable machinery, make better predictions, find workable solutions to potential problems, and evaluate the limitations of our current understanding. It's up to us to use our knowledge well.

We should remember that the full scope of applications of scientific discoveries is rarely apparent at first. Yet scientific advances can surreptitiously change our world, as well as our worldview. If applied well, they can yield tremendous benefits. Even many insights rooted in abstract theory—basic research that no one initially thought would ever have practical applications—has had a profound impact on our world.

Genetic studies, which today aim to treat cancer, are rooted in DNA research that initially was focused on purely theoretical questions. Medical tools such as MRIs grew out of our understanding of the atomic nucleus. Nuclear energy, which has been used for good and ill, emerged from knowledge of the structure of the atom. The electronics revolution grew out of transistor development, which developed from quantum physics. The Internet was a byproduct of the computer scientist Tim Berners-Lee's work at CERN—the particle physics accelerator center that now hosts the Large Hadron Collider—to improve communication and coordination among scientists in different nations,. GPS systems—ubiquitous today—incorporate Einstein's theory of relativity. No one even knew electricity would be important when it was first discovered, yet it is crucial to our current way of life.

When first starting his studies, the geologist Walter Alvarez, whose father was a Nobel-prize winning physicist, thought geology was routine compared to physics. Geologists were reconstructing relatively prosaic river and land patterns, while twentieth century physicists were radically changing the way people thought about the world and how it works. But as plate tectonics, stratigraphy, and geological evolution became better understood, oil reserves and mineralogical deposits were discovered and exploited. What started off as idle curiosity developed into tools for searching for oil and minerals. The transition began in the 18th century, but geology escalated in importance in the 20th. Geology provided great dividends to our world. It literally fueled the modern industrial complex—and with it, our economy and lifestyles—but many of our current environmental troubles too.

However, as the legacy of Alvarez and others demonstrates, not just industrial applications, but basic research goals as well, have fueled important advances in our understanding of geology. Connecting the meteoroid and the Solar System to a wider context—the structure of the galaxy—seems like the right progression in this expanding intellectual adventure of grasping the connections between our world and the surrounding Universe.

• • •

I like to think of the research this book describes as a continuation of the appreciation of other sciences that Alvarez's work precipitated with its intertwining of geology, chemistry, and physics. That dark matter might complete the set of known relationships reinforces this continuity. Not only can we use geology to understand a cosmic event, but maybe with a detailed understanding of the nature of dark matter will we come to understand the dynamics that threw a comet into our path in the first place.

Though most people's interest in meteoroids—astronomers and asteroid-mining-investors aside—derives from their potential consequences for life, the imminent danger presented by these flying objects is fairly minimal. Asteroids and comets are mostly in stable orbits and those that do deviate to hit the Earth are usually pretty small. Only rarely do big objects digress enough to leave the Solar System or strike the Earth. Based on the information I've presented, I hope you now have a better sense of what extraterrestrial bodies could hit in the future and how dangerous their impact is likely to be.

This book has explained the multiple threads of evidence establishing the one solid example of such a connection, which is the K-Pg extinction 66 million years in the past. In some global sense, we are all descendents of Chicxulub. It's a part of our history that we should want to understand. If true, the additional wrinkle presented in this book would mean that not only was dark matter responsible for irrevocably changing our world, but also that some of it played a crucial role in allowing for our existence. In this scenario, from the dinosaurs' perspective, dark matter was evil after all and the name scientists gave it was apt. But from a human perspective, this newly proposed type of dark matter was the instigator of one of the central accidents that changed the course of the Earth's development to the point where you could be sitting here reading this book.

In *Dark Matter and the Dinosaurs*, I've tried to give a taste of the nature of scientific investigations—how we pin down what we know and advance beyond to uncharted realms. On the other hand, the history of the Universe and of the Earth takes us on an exciting but challenging voyage into our past. If you think family history is difficult to trace—even when people are still around to tell the tale—think of the obstacles to extricating a past preserved only in inanimate rocks, many of which have eroded over time or been subducted back into the mantle below or the complexity of understanding how dark substances, which we can't even see, created structure.

Yet progress in science has revealed some of the remarkably intricate connections between the most fundamental matter's physical makeup and the features of the world that we see. Particles of dark matter collapsed to form galaxies, heavy elements created inside stars have been absorbed into life, the energy released by decays of radioactive nuclei deep inside the mantle have driven the movements of the Earth's crust that have created mountains. I find our ability to make progress in understanding these deeper connections in the Universe truly inspiring. Every time scientists have explored the boundaries of the known world, unanticipated discoveries have emerged.

Our world is rich—so rich that two of the most important questions particle physicists ask are, "Why this richness?" "How is all the matter we see related?" In my research, I am aware that what I am investigating may or may not ultimately connect directly to our experience of the world around us but hoping that, no matter what, the results will contribute to further progress. I concentrate on the task at hand, aware that anything that doesn't fit into our standard picture and calculations could either indicate an inadequate understanding of conventional models or it could be a sign of something new.

Dark matter and its role in the Universe's evolution are some of the most exciting scientific topics today. We will truly understand all forms of matter—as with any culturally complex society—only when we recognize and value the ways in which the diverse populations together contribute to the richness of our environment. The best way to advance our understanding will be to determine the most elegant and reliable table setting of matter that fits with observations. The dark matter proposal I've presented might be a thought experiment for now, but it's one that will be validated or invalidated by actual future measurements. Data and theoretical consistency together are the uncompromising arbiters of what is right.

The speculative influence on a comet for which this book is titled is not the only possible implication of the new type of dark matter we've proposed. A dark matter disk could affect the motions of stars, the makeup of dwarf galaxies, and the results of experiments and observations in laboratories and in space. Though understanding dark matter has been even more elusive in many respects than exploring the Earth and the Solar System, scientists are finding novel ways to track it down. The results will tell us about the make-up of our galaxy and of our Universe.

Our planet probably doesn't contain the only forms of life in the cosmos. But our existence required and still requires a universe and a planet with a number of remarkable properties. Forces we are only beginning to understand have been essential to how we got here. Understanding the galaxy and our origins within it yields a wider perspective on the felicitous accidents as well as the more predictable evolutionary processes that got us to this point. The amount that we already grasp is remarkable, as are the many more connections we aim to reveal. The progress of the last 50 years is nothing short of astonishing.

Though we are often faced with discouraging headlines and disappointingly cyclical patterns in world events, expanding scientific knowledge has the potential to enrich our lives and guide our actions in ways that preserve what we most value as we continue to advance. As research continues to uncover still more of the bridges that link our lives to our surroundings and our present to our past, we should appreciate the many features of our world that have been so long in the making and take care to use our acquired wisdom and technological advances well.

I always find it heartening to remember our amazing cosmological context. The petty squabbles of the world and short-term concerns shouldn't distract us from the enormous scope of what science can teach us about the world. The words I'm about to say might not always sound like the most practical advice. But look up. And around you. A fascinating Universe is out there for us to cultivate, cherish, and understand.

ACKNOWLEDGMENTS

This book was inspired by my research in physics, and by the many ideas in astronomy, geology, and biology to which the particular research I presented here have led me. Many scientists have contributed to my immersion in these subjects and I wish to thank all the physicists and astronomers who have shared their knowledge during this and other research. *Dark Matter and the Dinosaurs* also reflects my fascination with and excitement about our world, as well as my concerns about some of its directions. The shaping of these ideas owes a great debt to the many friends with whom I've had enlightening conversations over the years. I'd like to thank everyone who helped along the way.

I'd like to particularly thank all the many colleagues who have shared my scientific interest and especially my collaborators on various aspects of the disk dark matter research: JiJi Fan, Andrey Katz, Eric Kramer, Matthew McCullough, Matthew Reece, and Jakub Scholtz. Thanks also to Paul Davies for suggesting the connection that guides the ideas in this book and to Matthew Reece for collaborating on it. I was fortunate to have Matthew and Lubos Motl read an early draft and I thank them both for their thoughts and encouragement (though Lubos' ideas about some controversial issues made his encouragement somewhat selective . . .).

I also want to thank the physics and astronomy colleagues who looked over various chapters, including Laura Baudis, James Bullock, Bogdan Dobrescu, Doug Finkbeiner, Richard Gaitskell, Jakub Scholtz, and Tim Tait. Adam Brown's checking a nearly final version was also quite valuable. Comments on the science by Jo Bovi, Matthew Buckley, Sean Carroll, Chris Flynn, Lars Bergstrum, Ken Farley, Lars Hernquist, Johan Holmberg, Avi Loeb, Jonathan McDowell, Scott Tremaine, and Matt Walker are also reflected in the book's contents, as are the contributions of the astronomers who reviewed facts and provided insights—especially Francesca DeMeo, Dmitar Sasselov, and Maria Zuber. Special thanks are due to Martin Elvis and Chris Flynn who were very generous throughout with their time and input. Jerry Coyne, Nathan Mhyrvhold and especially Walter Alvarez, Andy Knoll, and David Kring provided valuable insider perspectives on extinctions and the K-Pg extinction in particular and their suggestions and corrections were invaluable. I am also grateful to Jose Juan Blanco, Asier Hilario, Miren Mendea, and Jon Urrestilla for arranging the visit to the K-Pg boundary in Spain.

But knowing the science is one thing. Writing a book is another. In addition to the insightful comments and thoughtful reading of my hard science colleagues, I was extremely fortunate to benefit from the support and wisdom of many friends with other interests. I thank Andi Machl for his time, support, and reliability throughout and for his frank assessments of when I was saying too much or too little. The high standards of Cormac McCarthy (and his muted manner of expressing disapproval) after a thoughtful reading also helped spur the project forward. The wisdom, advice, and encouragement of my friends Judith Donaugh, Maya Jasanoff, and Jen Sacks were very helpful in shaping the ideas and words in this book and Jen's insights also helped with the many pieces. The command of the English language of David Lewis and the editorial wisdom of Anna Christina Buchmann greatly contributed to the final product. I also wish to thank Jim Brooks, Richard Engel, Timothy Ferris, Milo Goodell, Tom Levenson, Howard Lutnick, Dana Randall, and Michael Snediker, who all made good points that contributed.

I am very grateful to my editor Hilary Redmon for her advice, encouragement, and patience during the completion of this book and her assistant Emma Janaskie for helping put the pieces together. Stuart Williams at Random House UK also provided valuable editorial insights. I also thank Dan Halpern and the staff at Ecco for their help and Alison Saltzberg for graciously working with me on the cover. The talented and thoughtful Rose Lincoln helped with a portrait photo, Gary Pikovsky provided new illustrations, Elisabeth Cheries, Robin Green, Emma Janaskie, Eric Kaplan, David Kring, Emily Lakdawalla, Tommy McCall, and Bill Prady helped with a few of the pictures, Kathleen Rocheleau was very skillful with references, and Elisabeth Cheries helped with proofreading. I thank Yaddo and my co-residents there for an enjoyable and productive residence and Marty and Sarah Flug for their hospitality at some important junctures and to Harvard for the time to work on this and for a productive physics environment. Many thanks also to my agent Andrew Wylie for helping launch this project, and to both Andrew and Sarah Chalfant for encouragement on a draft. Thanks also to the rest of the Wylie Agency team, including James Pullen and Kristina Moore for smoothing things along the way.

Special gratitude is owed to Jeff Goodell, who repeatedly shared the insights of a skillful writer who understands the momentum of a good story as well as the wisdom of knowing how to convey it, and whose curiosity I was happy to share. Thanks also to his family and to my family and friends for their invaluable curiosity, interest, and encouragement.

Finally, I wish to thank my parents, who sadly won't get to share in this book, but whose influence I trust is present throughout in the perspective that guided it. I want to thank them for inspiring me to believe goals are attainable, including ambitious undertakings like this one.

LIST OF ILLUSTRATIONS

SUPPLEMENTARY READING

Below is a compilation of some of the papers and that books that I have found interesting and useful. This is not intended as a comprehensive survey. A lot of the entries are dedicated to more controversial subjects, but I did also include some review articles and a few key papers. Basic topics are also covered in textbooks and also Wikipedia, which knowledgeable enthusiasts keep fairly up-to-date on many science topics.

CHAPTERS 1 AND 2

Bergström, Lars. "Non-Baryonic Dark Matter: Observational Evidence and Detection Methods." *Reports on Progress in Physics* 63.5 (2000): 793–841.

Bertone, Gianfranco, Dan Hooper, and Joseph Silk. "Particle Dark Matter: Evidence, Candidates and Constraints." *Physics Reports* 405.5-6 (2005): 279–390.

Copi, C J, D N Schramm, and M S Turner. "Big-Bang Nucleosynthesis and the Baryon Density of the Universe." *Science* 267.5195 (1995): 192–9.

Freese, Katherine. *The Cosmic Cocktail: Three Parts Dark Matter.* Princeton University Press, 2014.

Garrett, Katherine, and Gintaras Duda. "Dark Matter: A Primer." *Advances in Astronomy* 2011 (2011): 1–22.

Gelmini, Graciela B. *TASI 2014 Lectures: The Hunt for Dark Matter.* (2015). http://arxiv.org/abs/1502.01320.

Lundmark, Knut. Lund Medd. 1 No125 = VJS 65, p. 275 (1930)

Olive, Keith A. "TASI lectures on dark matter." *arXiv preprint astro-ph/030 1505*(2003).

Panek, Richard. The 4 Percent Universe: Dark Matter, Dark Energy, and the Race to Discover the Rest of Reality: Mariner Books, 2011.

Peter, Annika HG. "Dark Matter: A Brief Review." *Frank N. Bash Symposium 2011: New Horizons in Astronomy.* Ed. Sarah Salviander, Joel Green, and Andreas Pawlik. University of Texas at Austin, 2012.

Profumo, Stefano. "TASI 2012 Lectures on Astrophysical Probes of Dark Matter." (2013): 41.

Rubin, V. C., N. Thonnard, and Jr. Ford, W. K. "Rotational Properties of 21 SC Galaxies with a Large Range of Luminosities and Radii, from NGC 4605 /R = 4kpc/ to UGC 2885 /R = 122 Kpc/." *The Astrophysical Journal* 238 (1980): 471–487.

Rubin, Vera C., and Jr. Ford, W. Kent. "Rotation of the Andromeda Nebula from a Spectroscopic Survey of Emission Regions." *The Astrophysical Journal* 159 (1970): 379–403.

Sahni, Varun. "Dark Matter and Dark Energy." *Physics of the Early Universe.* Springer Berlin Heidelberg, 2005. 141–179.

Strigari, Louis E. "Galactic Searches for Dark Matter." *Physics Reports* 531.1 (2013): 1–88.

Trimble, V. "Existence and Nature of Dark Matter in the Universe." *Annual Review of Astronomy and Astrophysics* 25 (1987): 425–472.

Zwicky, F. "Die Rotverschiebung von Extragalaktischen Nebeln." *Helvetica Physica Acta* 6 (1933): 110–127.

Zwicky, F. "On the Masses of Nebulae and of Clusters of Nebulae." *The Astrophysical Journal* 86 (1937): 217.

CHAPTER 3

Humboldt, Alexander von. Kosmos: A General Survey of Physical Phenomena of the Universe, Volume 1. H. Baillière, 1845.

CHAPTER 4

Baumann, Daniel. "TASI Lectures on Inflation." (2009). http://arxiv.org/abs/0907.5424.

Boggess, N. W. et al. "The COBE Mission—Its Design and Performance Two Years after Launch." *The Astrophysical Journal* 397 (1992): 420–429.

Freeman, Ken, and Geoff McNamara. *In Search of Dark Matter.* Springer, 2006.

Guth, Alan H. The Inflationary Universe: The Quest for a New Theory of Cosmic Origins. Perseus Books, 1997.

Hinshaw, G. et al. "Five-Year Wilkinson Microwave Anisotropy Probe Observations: Data Processing, Sky Maps, and Basic Results." *The Astrophysical Journal Supplement Series* 180.2 (2009): 225–245.

Kamionkowski, Marc, Arthur Kosowsky, and Albert Stebbins. "A Probe of Primordial Gravity Waves and Vorticity." *Physical Review Letters* 78 (1997): 2058–2061.

Komatsu, E. et al. "Five-Year Wilkinson Microwave Anisotropy Probe Observations: Cosmological Interpretation." *The Astrophysical Journal Supplement Series* 180.2 (2009): 330–376.

Kowalski, M. et al. "Improved Cosmological Constraints from New, Old, and Combined Supernova Data Sets." *The Astrophysical Journal* 686.2 (2008): 749–778.

Leitch, E. M. et al. "Degree Angular Scale Interferometer 3 Year Cosmic Microwave Background Polarization Results." *The Astrophysical Journal* 624.1 (2005): 10–20.

Penzias, A. A., and R. W. Wilson. "A Measurement of Excess Antenna Temperature at 4080 Mc/s." *The Astrophysical Journal* 142 (1965): 419–421.

Seljak, Uroš, and Matias Zaldarriaga. "Signature of Gravity Waves in the Polarization of the Microwave Background." *Physical Review Letters* 78.11 (1997): 2054–2057.

Weinberg, Steven. The First Three Minutes: A Modern View of the Origin of the Universe. Basic Books, 1993.

CHAPTER 5

Binney, J., and S. Tremaine. *Galactic Dynamics*. Princeton University Press, 2008.

Davis, M. et al. "The Evolution of Large-Scale Structure in a Universe Dominated by Cold Dark Matter." *The Astrophysical Journal* 292 (1985): 371–394.

"Hubble Maps the Cosmic Web of 'Clumpy' Dark Matter in 3-D." (7 January 2007). http://hubblesite.org/newscenter/archive/releases/2007/01/image/a/grav/.

Kaehler, R., O. Hahn, and T. Abel. "A Novel Approach to Visualizing Dark Matter Simulations." *IEEE Transactions on Visualization and Computer Graphics* 18.12 (2012): 2078–2087.

Loeb, Abraham. *How Did the First Stars and Galaxies Form?* Princeton University Press, 2010.

Loeb, Abraham, and Steven R. Furlanetto. *The First Galaxies in the Universe*. Princeton University Press, 2013.

Massey, Richard et al. "Dark Matter Maps Reveal Cosmic Scaffolding." *Nature* 445.7125 (2007): 286–90.

Mo, Houjun, Frank van den Bosch, and Simon White. *Galaxy Formation and Evolution*. Cambridge University Press, 2010.

Papastergis, Emmanouil et al. "A Direct Measurement of the Baryonic Mass Function of Galaxies & Implications for the Galactic Baryon Fraction." *Astrophysical Journal* 259.2 (2012): 138.

Springel, Volker et al. "Simulations of the Formation, Evolution and Clustering of Galaxies and Quasars." *Nature* 435.7042 (2005): 629–36.

CHAPTER 6

Blitzer, Jonathan. "The Age of Asteroids." *New Yorker*. (2014). http://www.newyorker.com/tech/elements/age-asteroids.

DeMeo, F E, and B Carry. "Solar System Evolution from Compositional Mapping of the Asteroid Belt." *Nature* 505 (2014): 629–34.

Kleine, Thorsten et al. "Hf–W Chronology of the Accretion and Early Evolution of Asteroids and Terrestrial Planets." *Geochimica et Cosmochimica Acta* 73.17 (2009): 5150–5188.

Lissauer, Jack J., and Imke de Pater. *Fundamental Planetary Science: Physics, Chemistry and Habitability*. Cambridge University Press, 2013.

Rubin, Alan E., and Jeffrey N. Grossman. "Meteorite and Meteoroid: New Comprehensive Definitions." *Meteoritics and Planetary Science* (2010): 114-122.

CHAPTER 7

Bailey, M. E., and C. R. Stagg. "Cratering Constraints on the Inner Oort Cloud: Steady-State Models." *Monthly Notices of the Royal Astronomical Society* 235.1 (1988): 1–32.

"Europe's Comet Chaser." *European Space Agency*. (2014). http://www.esa.int/Our_Activities/Space_Science/Rosetta/Europe_s_comet_chaser.

Gladman, B. "The Kuiper Belt and the Solar System's Comet Disk." *Science* 307.5706 (2005): 71–75.

Gladman, B., B. G. Marsden, and C. Vanlaerhoven. "Nomenclature in the

Outer Solar System." *The Solar System Beyond Neptune*. Ed. M. A. Barucci et al. University of Arizona Press, 2008. 43–57.

Gomes, Rodney. "Planetary Science: Conveyed to the Kuiper Belt." *Nature* 426.6965 (2003): 393–5.

Iorio, L. "Perspectives on Effectively Constraining the Location of a Massive Trans-Plutonian Object with the New Horizons Spacecraft: A Sensitivity Analysis." *Celestial Mechanics and Dynamical Astronomy* 116.4 (2013): 357–366.

Morbidelli, A, and H F Levison. "Planetary Science: Kuiper-Belt Interlopers." *Nature* 422.6927 (2003): 30–1.

Olson, RJM. "Much Ado about Giotto's Comet." *Quarterly Journal of the Royal Astronomical Society* 35.1 (1994): 145.

Robinson, Howard. The Great Comet of 1680: A Study in the History of Rationalism. Press of the Northfield News, 1916.

Walsh, Colleen. "The Building Blocks of Planets." *Harvard Gazette* 12 Sept. 2013.

CHAPTER 8

Francis, Matthew. "The Solar System Boundary and the Week in Review (September 8-14)." *Bowler Hat Science*. (2013). http://bowlerhatscience.org/2013/09/14/the-solar-system-boundary-and-the-week-in-review-september-8-14/.

McComas, David. "What Is the Edge of the Solar System Like?—NOVA Next | PBS." (2013). http://www.pbs.org/wgbh/nova/next/space/voyager-ibex-and-the-edge-of-the-solar-system/.

CHAPTER 9

Gehrels, T., ed. *Hazards Due to Comets and Asteroids*. University of Arizona Press, 1995.

The Earth Institute, "The Growing Urbanization of the World," Columbia University, New York, 2005.

"IAU Minor Planet Center." (2015). http://www.minorplanetcenter.net/.

Kring, David A., and Mark Boslough. "Chelyabinsk: Portrait of an Asteroid Airburst." *Physics Today* 67.9 (2014): 32–37.

Levison, Harold F et al. "The Mass Disruption of Oort Cloud Comets." *Science* 296.5576 (2002): 2212–5.

Marvin, U. B. "Siena, 1794: History's Most Consequential Meteorite Fall." *Meteoritics* 30.5 (1995): 540.

Marvin, Ursula B. "Ernst Florens Friedrich Chladni (1756-1827) and the

Origins of Modern Meteorite Research." *Meteoritics & Planetary Science* 31.5 (1996): 545–588.

Marvin, Ursula B. "Meteorites in History: An Overview from the Renaissance to the 20th Century." *Geological Society, London, Special Publications* 256.1 (2006): 15–71.

Marvin, Ursula B., and Mario L. Cosmo. "Domenico Troili (1766): 'The True Cause of the Fall of a Stone in Albereto Is a Subterranean Explosion That Hurled the Stone Skyward.'" *Meteoritics & Planetary Science* 37.12 (2002): 1857–1864.

"Meteorites, Impacts, & Mass Extinction." (2014). http://www.tulane.edu/~sanelson/Natural_Disasters/impacts.htm.

National Research Council. Defending Planet Earth: Near-Earth Object Surveys and Hazard Mitigation Strategies. National Academies Press, 2010.

Nield, Ted. Incoming! Or, Why We Should Stop Worrying and Learn to Love the Meteorite. Granta Books, 2012.

Shapiro, Irwin I. et al., with National Research Council. "Defending Planet Earth: Near-Earth Object Surveys and Hazard Mitigation Strategies." 2010: 149.

Tagliaferri, E. et al. "Analysis of the Marshall Islands Fireball of February 1, 1994." *Earth, Moon, and Planets* 68.1-3 (1995): 563–572.

CHAPTER 10

"Astronomy: Collision History Written in Rock." *Nature* 512.7515 (2014): 350.

Barringer, D. M. "Coon Mountain and Its Crater." Proceedings of the Academy of Natural Sciences of Philadelphia, Vol. 57. 1905. 861–886.

"Earth Impact Database." http://www.passc.net/EarthImpactDatabase/.

Grieve, R A F. "Terrestrial Impact Structures." *Annual Review of Earth and Planetary Sciences* 15 (1987): 245–270.

Kring, David A. Guidebook to the Geology of Barringer Meteorite Crater, Arizona. Lunar and Planetary Institute, 2007.

Tilghman, B. C. "Coon Butte, Arizona." Proceedings of the Academy of Natural Sciences of Philadelphia, v. 57 (1905). 887–914.

CHAPTER 11

Bambach, Richard K. "Phanerozoic Biodiversity Mass Extinctions." *Annual Review of Earth and Planetary Sciences* 34.1 (2006): 127–155.

Bambach, Richard K., Andrew H. Knoll, and Steve C. Wang. "Origination, Extinction, and Mass Depletions of Marine Diversity." *Paleobiology* 30.4 (2004): 522–542.

Barnosky, Anthony D. Dodging Extinction: Power, Food, Money, and the Future of Life on Earth. University of California Press, 2014.

Barnosky, Anthony D et al. "Has the Earth's Sixth Mass Extinction Already Arrived?" *Nature* 471 (2011): 51–57.

Carpenter, Kenneth. Eggs, Nests, and Baby Dinosaurs: A Look at Dinosaur Reproduction. Indiana University Press, 1999.

Eldredge, Niles. Reinventing Darwin: The Great Debate at the High Table of Evolutionary Theory. Wiley, 1995.

Jablonski, David. "Mass Extinctions and Macroevolution." *Paleobiology* 31.sp5 (2005): 192–210.

Kelley, S. "The Geochronology of Large Igneous Provinces, Terrestrial Impact Craters, and Their Relationship to Mass Extinctions on Earth." *Journal of the Geological Society* 164.5 (2007): 923–936.

Kidwell, Susan M. "Shell Composition Has No Net Impact on Large-Scale Evolutionary Patterns in Mollusks." *Science* 307 (2005): 914–917.

Kolbert, Elizabeth. *The Sixth Extinction: An Unnatural History*. Henry Holt & Co., 2014.

Kurtén, Björn. *Age Groups in Fossil Mammals*. Helsinki: Societas scientiarum Fennica, 1953.

Lawton, John H., and Robert May, eds. *Extinction Rates*. Oxford University Press, 1995.

MacLeod, Norman. The Great Extinctions: What Causes Them and How They Shape Life. Firefly Books, 2013.

"Modern Extinction Estimates." (2015). http://rainforests.mongabay.com/09x_table.htm.

Newell, Norman D. "Revolutions in the History of Life." *Geological Society of America Special Papers* 89. Geological Society of America, 1967. 63–92.

Pimm, S. L. et al. "The Biodiversity of Species and Their Rates of Extinction, Distribution, and Protection." *Science* 344 (2014): 1246752.

Rothman, Daniel H et al. "Methanogenic Burst in the End-Permian Carbon Cycle." *Proceedings of the National Academy of Sciences of the United States of America* 111.15 (2014): 5462–7.

Sanders, Robert. "Has the Earth's Sixth Mass Extinction Already Arrived?" *UC Berkeley News Center* 2 Mar. 2011.

Schindel, David E. "Microstratigraphic Sampling and the Limits of Paleontologic Resolution." *Paleobiology* 6.4 (1980): 408–426.

Sepkoski, J. J. "Phanerozoic Overview of Mass Extinction." *Patterns and Processes in the History of Life*. Ed. D. M. Raup and D. Jablonski. Springer Berlin Heidelberg, 1986. 277–295.

Valentine, James W. "How Good Was the Fossil Record? Clues from the California Pleistocene." *Paleobiology* 15.2 (1989): 83–94.

Wilson, Edward O. *The Future of Life*. 1st ed. Vintage Books, 2003.

CHAPTER 12

Alvarez, Walter. *T. Rex and the Crater of Doom*. Princeton University Press, 2008.

Alvarez, L. W. et al. "Extraterrestrial Cause for the Cretaceous-Tertiary Extinction." *Science* 208.4448 (1980): 1095–108.

Caldwell, Brady. "The K-T Event: A Terrestrial or Extraterrestrial Cause for Dinosaur Extinction?" *Essay in Palaeontology 5p* (2007).

Choi, Charles Q. "Asteroid Impact That Killed the Dinosaurs: New Evidence." (2013). *http://www.livescience.com/26933-chicxulub-cosmic-impact-dinosaurs.html*.

Frankel, Charles. The End of the Dinosaurs: Chicxulub Crater and Mass Extinctions. Cambridge University Press, 1999.

Kring, David A. et al. "Impact Lithologies and Their Emplacement in the Chicxulub Impact Crater: Initial Results from the Chicxulub Scientific Drilling Project, Yaxcopoil, Mexico." *Meteoritics & Planetary Science* 39.6 (2004): 879–897.

Kring, David A. et al. "The Chicxulub Impact Event and Its Environmental Consequences at the Cretaceous–Tertiary Boundary." *Palaeogeography, Palaeoclimatology, Palaeoecology* 255.1-2 (2007): 4–21.

Moore, J. R., and M. Sharma. "The K-Pg Impactor Was Likely a High Velocity Comet." *44th Lunar and Planetary Conference; Paper #2431*. 2013.

Ravizza, G, and B Peucker-Ehrenbrink. "Chemostratigraphic Evidence of Deccan Volcanism from the Marine Osmium Isotope Record." *Science* 302.5649 (2003): 1392–5.

Sanders, Robert. "New Evidence Comet or Asteroid Impact Was Last Straw for Dinosaurs." *UC Berkeley News Center* 7 Feb. 2013.

Schulte, Peter et al. "The Chicxulub Asteroid Impact and Mass Extinction at the Cretaceous-Paleogene Boundary." *Science* 327.5970 (2010): 1214–8.

"What Killed the Dinosaurs? The Great Mystery - Background." http://www.ucmp.berkeley.edu/diapsids/extinction.html.

"What Killed the Dinosaurs? The Great Mystery - Invalid Hypotheses http://www.ucmp.berkeley.edu/diapsids/extincthypo.html.

Zahnle, K, and D Grinspoon. "Comet Dust as a Source of Amino Acids at the Cretaceous/Tertiary Boundary." *Nature* 348.6297 (1990): 157–60.

CHAPTER 13

American Chemical Society. "New Evidence That Comets Deposited Building Blocks of Life on Primordial Earth." *Science Daily* (2012): 27 March. www.sciencedaily.com/releases/2012/03/120327215607.htm.

"Astronomy: Comets Forge Organic Molecules." *Nature* 512.7514 (2014): 234–235.

Durand-Manterola, Hector Javier, and Guadalupe Cordero-Tercero. "Assessments of the Energy, Mass and Size of the Chicxulub Impactor." (2014). arXiv:1403:6391.

Elvis, Martin. "Astronomy: Cosmic Triangles and Black-Hole Masses." *Nature* 515.7528 (2014): 498–499.

Elvis, Martin. "How Many Ore-Bearing Asteroids?" *Planetary and Space Science* 91 (2014): 20–26.

Elvis, Martin, and Thomas Esty. "How Many Assay Probes to Find One Ore-Bearing Asteroid?" *Acta Astronautica* 96 (2014): 227–231.

Knoll, Andrew H. Life on a Young Planet: The First Three Billion Years of Evolution on Earth. Princeton University Press, 2003.

Livio, Mario, Neill Reid, and William Sparks, eds. Astrophysics of Life: Proceedings of the Space Telescope Science Institute Symposium, Held in Baltimore, Maryland May 6–9, 2002. Cambridge University Press, 2005.

Melott, Adrian L., and Brian C. Thomas. "Astrophysical Ionizing Radiation and Earth: A Brief Review and Census of Intermittent Intense Sources." *Astrobiology* 11.4 (2011): 343–361.

Poladian, Charles. "Comets Or Meteorites Crashing Into A Planet Could Produce Amino Acids, 'Building Blocks Of Life.'" *International Business Times* 15 Sept. 2013.

Rothery, David A., Mark A. Sephton, and Iain Gilmour, Eds. *An Introduction to Astrobiology.* 2nd ed. Cambridge University Press, 2011.

Steigerwald, Bill. "Amino Acids in Meteorites Provide a Clue to How Life Turned Left." 2012. http://scitechdaily.com/amino-acids-in-meteorites-provide-a-clue-to-how-life-turned-left/.

CHAPTER 14

Alvarez, Walter, and Richard A. Muller. "Evidence from Crater Ages for Periodic Impacts on the Earth." *Nature* 308 (1984): 718–720.

Bailer-Jones, C. A. L. "Bayesian Time Series Analysis of Terrestrial Impact Cratering." *Monthly Notices of the Royal Astronomical Society* 416.2 (2011): 1163–1180.

Bailer-Jones, C. A. L. "Evidence for a Variation - but No Periodicity - in the Terrestrial Impact Cratering Rate." *EPSC-DPS Joint Meeting 2011* (2011): 153.

Bailer-Jones, C. A. L. "The Evidence for and against Astronomical Impacts on Climate Change and Mass Extinctions: A Review." *International Journal of Astrobiology* 8.3 (2009): 213.

Chang, Heon-Young, and Hong-Kyu Moon. "Time-Series Analysis of Terrestrial Impact Crater Records." *Publications of the Astronomical Society of Japan* 57.3 (2005): 487–495.

Connor, E. F. "Time Series Analysis of the Fossil Record." *Patterns and Processes in the History of Life*. Springer Berlin Heidelberg, 1986. 119–147.

Feulner, Georg. "Limits to Biodiversity Cycles from a Unified Model of Mass-Extinction Events." *International Journal of Astrobiology* 10.02 (2011): 123–129.

Fox, William T. "Harmonic Analysis of Periodic Extinctions." *Paleobiology* 13.3 (1987): 257–271.

Grieve, R. A. F. et al. "Detecting a Periodic Signal in the Terrestrial Cratering Record." *Lunar and Planetary Science Conference* (1988): 375–382.

Grieve, R. A. F., and D. A. Kring. "Geologic Record of Destructive Impact Events on Earth." *Comet/Asteroid Impacts and Human Society: An Interdisciplinary Approach*. Ed. Peter T. Bobrowsky and Hans Rickman. Springer Berlin Heidelberg, 2007. 3–24.

Grieve, Richard A. F. "Terrestrial Impact: The Record in the Rocks*." *Meteoritics* 26.3 (1991): 175–194.

Grieve, Richard A. F., and Eugene M. Shoemaker. "The Record of Past Impacts on Earth." *Hazards Due to Comets and Asteroids*. Ed. T. Gehrels. University of Arizona Press, 1994. 417–462.

Heisler, Julia, and Scott Tremaine. "How Dating Uncertainties Affect the Detection of Periodicity in Extinctions and Craters." *Icarus* 77.1 (1989): 213–219.

Heisler, Julia, Scott Tremaine, and Charles Alcock. "The Frequency and Intensity of Comet Showers from the Oort Cloud." *Icarus* 70.2 (1987): 269–288.

Jetsu, L., and J. Pelt. "Spurious Periods in the Terrestrial Impact Crater Record." *Astronomy and Astrophysics* 353 (2000): 409–418.

Lieberman, Bruce S. "Whilst This Planet Has Gone Cycling On: What Role for Periodic Astronomical Phenomena in Large-Scale Patterns in the History of Life?" *Earth and Life: Global Biodiversity, Extinction Intervals and Biogeographic Perturbations Through Time.* Springer Netherlands, 2012. 37–50.

Lyytinen, J. et al. "Detection of Real Periodicity in the Terrestrial Impact Crater Record: Quantity and Quality Requirements." *Astronomy and Astrophysics* 499.2 (2009): 601–613.

Melott, Adrian L. et al. "A ~60 Myr Periodicity Is Common to Marine-87Sr/86Sr, Fossil Biodiversity, and Large-Scale Sedimentation: What Does the Periodicity Reflect?" *Journal of Geology* 120 (2012): 217–226.

Melott, Adrian L., and Richard K. Bambach. "A Ubiquitous ~62-Myr Periodic Fluctuation Superimposed on General Trends in Fossil Biodiversity. I. Documentation." *Paleobiology* 37.1 (2011): 92–112.

Melott, Adrian L., and Richard K. Bambach. "Analysis of Periodicity of Extinction Using the 2012 Geological Timescale." *Paleobiology* 40.2 (2014): 177–196.

Melott, Adrian L., and Richard K. Bambach. "Do Periodicities in Extinction - with Possible Astronomical Connections - Survive a Revision of the Geological Timescale?" *The Astrophysical Journal* 773.1 (2013): 1–5.

Noma, Elliot, and Arnold L. Glass. "Mass Extinction Pattern: Result of Chance." *Geological Magazine* 124.4 (1987): 319–322.

Quinn, James F. "On the Statistical Detection of Cycles in Extinctions in the Marine Fossil Record." *Paleobiology* 13.4 (1987): 465–478.

Raup, D. M., and J J Sepkoski. "Mass Extinctions in the Marine Fossil Record." *Science* 215.4539 (1982): 1501–3.

Raup, D. M., and J. J. Sepkoski. "Periodicity of Extinctions in the Geologic Past." *Proceedings of the National Academy of Sciences* 81.3 (1984): 801–805.

Raup, D., and J. Sepkoski. "Periodic Extinction of Families and Genera." *Science* 231.4740 (1986): 833–836.

Stigler, S M, and M J Wagner. "A Substantial Bias in Nonparametric Tests for Periodicity in Geophysical Data." *Science* 238.4829 (1987): 940–5.

Stothers, Richard B. "Structure and Dating Errors in the Geologic Time Scale and Periodicity in Mass Extinctions." *Geophysical Research Letters* 16.2 (1989): 119–122.

Stothers, Richard B. "The Period Dichotomy in Terrestrial Impact Crater Ages." *Monthly Notices of the Royal Astronomical Society* 365.1 (2006): 178–180.

Trefil, J.S., and D.M. Raup. "Numerical Simulations and the Problem of Periodicity in the Cratering Record." *Earth and Planetary Science Letters* 82.1-2 (1987): 159–164.

Yabushita, S. "A Statistical Test of Correlations and Periodicities in the Geological Records." *Celestial Mechanics and Dynamical Astronomy* 69.1-2 31–48.

Yabushita, S. "Are Cratering and Probably Related Geological Records Periodic?" *Earth, Moon and Planets* 72.1-3 (1996): 343–356.

Yabushita, S. "On the Periodicity Hypothesis of the Ages of Large Impact Craters." *Monthly Notices of the Royal Astronomical Society* 334.2 (2002): 369–373.

Yabushita, S. "Periodicity and Decay of Craters over the Past 600 Myr." *Earth, Moon and Planets* 58.1 (1992): 57–63.

Yabushita, S. "Statistical Tests of a Periodicity Hypothesis for Crater Formation Rate-II." *Monthly Notices of the Royal Astronomical Society* 279.3 (1996): 727–732.

CHAPTER 15

Davis, Marc, Piet Hut, and Richard A. Muller. "Extinction of Species by Periodic Comet Showers." *Nature* 308.5961 (1984): 715–717.

Filipovic, M. D. et al. "Mass Extinction and the Structure of the Milky Way." *Serbian Astronomical Journal* 87 (2013): 43–52.

Grieve, Richard A. F., and Lauri J. Pesonen. "Terrestrial Impact Craters: Their Spatial and Temporal Distribution and Impacting Bodies." *Earth, Moon and Planets* 72.1-3 (1996): 357–376.

Heisler, Julia, Scott Tremaine, and Charles Alcock. "The Frequency and Intensity of Comet Showers from the Oort Cloud." *Icarus* 70.2 (1987): 269–288.

Matese, J. "Periodic Modulation of the Oort Cloud Comet Flux by the Adiabatically Changing Galactic Tide." *Icarus* 116.2 (1995): 255–268.

Matese, J.J., K. A. Innanen, and M. J. Valtonen. "Variable Oort Cloud Flux due to the Galactic Tide." *Collisional Processes in the Solar System*. Ed.

Mikhail Marov and Hans Rickman. Kluwer Academic Publishers, 2001. 91–102.

Melott, Adrian L., and Richard K. Bambach. "Nemesis Reconsidered." *Monthly Notices of the Royal Astronomical Society: Letters* 407.1 (2010): L99–L102.

Napier, W. M. "Evidence for Cometary Bombardment Episodes." *Monthly Notices of the Royal Astronomical Society* 366.3 (2006): 977–982.

Nurmi, P., M. J. Valtonen, and J. Q. Zheng. "Periodic Variation of Oort Cloud Flux and Cometary Impacts on the Earth and Jupiter." *Monthly Notices of the Royal Astronomical Society* 327.4 (2001): 1367–1376.

Rampino, M. R. "Disc Dark Matter in the Galaxy and Potential Cycles of Extraterrestrial Impacts, Mass Extinctions and Geological Events." *Monthly Notices of the Royal Astronomical Society* 448.2 (2015): 1816–1820.

Rampino, M. R. "Galactic Triggering of Periodic Comet Showers." *Collisional Processes in the Solar System*. Ed. Mikhail Ya Marov and Hans Rickman. Kluwer Academic Publishers, 2001. 103–120.

Rampino, Michael, Bruce M. Haggerty, and Thomas C. Pagano. "A Unified Theory of Impact Crises and Mass Extinctions: Quantitative Tests." *Annals of the New York Academy of Sciences* 822.1 (1997): 403–431.

Rampino, Michael R., and Richard B. Stothers. "Terrestrial Mass Extinctions, Cometary Impacts and the Sun's Motion Perpendicular to the Galactic Plane." *Nature* 308 (1984): 709–712.

Schwartz, Richard D., and Philip B. James. "Periodic Mass Extinctions and the Sun's Oscillation about the Galactic Plane." *Nature* 308.5961 (1984): 712–713.

Shoemaker, Eugene M. "Impact Cratering Through Geologic Time." *Journal of the Royal Astronomical Society of Canada* 92 (1998): 297–309.

Smoluchowski, R., J.M. Bahcall, and M.S. Matthews. *Galaxy and the Solar System*. University of Arizona Press, 1986.

Stothers, R. B. "Galactic Disc Dark Matter, Terrestrial Impact Cratering and the Law of Large Numbers." *Monthly Notices of the Royal Astronomical Society* 300.4 (1998): 1098–1104.

Swindle, T. D., D. A. Kring, and J. R. Weirich. "40Ar/39Ar Ages of Impacts Involving Ordinary Chondrite Meteorites." *Geological Society, London, Special Publications* 378.1 (2013): 333–347.

Torbett, Michael V. "Injection of Oort Cloud Comets to the Inner Solar System by Galactic Tidal Fields." *Monthly Notices of the Royal Astronomical Society* 223 (1986): 885–895.

Wickramasinghe, J. T., and W. M. Napier. "Impact Cratering and the Oort Cloud." *Monthly Notices of the Royal Astronomical Society* 387.1 (2008): 153–157.

Whitmire, Daniel P., and Albert A. Jackson. "Are Periodic Mass Extinctions Driven by a Distant Solar Companion?" *Nature* 308.5961 (1984): 713–715.

Whitmire, Daniel P., and John J. Matese. "Periodic Comet Showers and Planet X." *Nature* 313.5997 (1985): 36–38.

Wickramasinghe, J. T., and W. M. Napier. "Impact Cratering and the Oort Cloud." *Monthly Notices of the Royal Astronomical Society* 387.1 (2008): 153–157.

CHAPTERS 16 AND 17

Ahmed, Z et al. "Dark Matter Search Results from the CDMS II Experiment." *Science* 327.5973 (2010): 1619–21.

Akerib, D. S. et al. "First Results from the LUX Dark Matter Experiment at the Sanford Underground Research Facility." *Physical Review Letters* 112.9 (2014): 091303.

Aprile, E. et al. "First Dark Matter Results from the XENON100 Experiment." *Physical Review Letters* 105.13 (2010).

Bergstrom, Lars. "Saas-Fee Lecture Notes: Multi-Messenger Astronomy and Dark Matter." (2012): 105.

Bertone, Gianfranco. "The Moment of Truth for WIMP Dark Matter." *Nature* 468.7322 (2010): 389–393.

Bertone, Gianfranco. *Particle Dark Matter: Observations, Models and Searches*. Cambridge University Press, 2010.

Bertone, Gianfranco, and David Merritt. "Dark Matter Dynamics and Indirect Detection." *Modern Physics Letters A* 20.14 (2005): 1021–1036.

Buckley, Matthew R., and Lisa Randall. "Xogenesis." *Journal of High Energy Physics* 9 (2011).

Cline, David B. "The Search for Dark Matter." *Scientific American* 288.3 (2003): 50–59.

Cohen, Timothy et al. "Asymmetric Dark Matter from a GeV Hidden Sector." *Physical Review D* 82.5 (2010).

Cohen, Timothy, and Kathryn M. Zurek. "Leptophilic Dark Matter from the Lepton Asymmetry." *Physical Review Letters* 104.10 (2010).

Cui, Yanou, Lisa Randall, and Brian Shuve. "A WIMPy Baryogenesis Miracle." *Journal of High Energy Physics* 4 (2012): 75.

Cui, Yanou, Lisa Randall, and Brian Shuve. "Emergent Dark Matter, Baryon, and Lepton Numbers." *Journal of High Energy Physics* 2011.8 (2011): 73.

Davoudiasl, Hooman et al. "Unified Origin for Baryonic Visible Matter and Antibaryonic Dark Matter." *Physical Review Letters* 105.21 (2010).

Drukier, Andrzej K., Katherine Freese, and David N. Spergel. "Detecting cold dark-matter candidates." *Physical Review D* 33.12 (1986): 3495.

Freeman, Ken, and Geoff McNamara. *In Search of Dark Matter.* Springer, 2006.

Gaitskell, Richard J. "Direct Detection of Dark Matter." *Annual Review of Nuclear and Particle Science* 54.1 (2004): 315–359.

Hooper, Dan, John March-Russell, and Stephen M. West. "Asymmetric Sneutrino Dark Matter and the Omega(b) / Omega(DM) Puzzle." *Physics Letters B* 605.3-4 (2005): 228–236.

Jungman, Gerard, Marc Kamionkowski, and Kim Griest. "Supersymmetric Dark Matter." *Physics Reports* 267.5-6 (1996): 195–373.

Kaplan, David B. "Single Explanation for Both Baryon and Dark Matter Densities." *Physical Review Letters* 68.6 (1992): 741–743.

Kaplan, David E., Markus A. Luty, and Kathryn M. Zurek. "Asymmetric Dark Matter." *Physical Review D* 79.11 (2009).

Napier, W. M. "Evidence for Cometary Bombardment Episodes." *Monthly Notices of the Royal Astronomical Society* 366.3 (2006): 977–982.

"Neutralino Dark Matter." http://www.picassoexperiment.ca/dm_neutralino.php.

Preskill, John, Mark B. Wise, and Frank Wilczek. "Cosmology of the Invisible Axion." *Physics Letters B* 120.1-3 (1983): 127–132.

Profumo, Stefano. "TASI 2012 Lectures on Astrophysical Probes of Dark Matter." (2013): 41.

Shelton, Jessie, and Kathryn M. Zurek. "Darkogenesis: A Baryon Asymmetry from the Dark Matter Sector." *Physical Review D* 82.12 (2010): 123512.

Thomas, Scott. "Baryons and Dark Matter from the Late Decay of a Supersymmetric Condensate." *Physics Letters B* 356.2-3 (1995): 256–263.

Turner, Michael S., and Frank Wilczek. "Inflationary Axion Cosmology." *Physical Review Letters* 66.1 (1991): 5–8.

Weinberg, Steven. "A New Light Boson?" *Physical Review Letters* 40.4 (1978): 223–226.

Wilczek, F. "Problem of Strong P and T Invariance in the Presence of Instantons." *Physical Review Letters* 40.5 (1978): 279–282.

CHAPTER 18

Ackerman, Lotty et al. "Dark Matter and Dark Radiation." *Physical Review D* 79.2 (2009): 023519.

Bovy, Jo, Hans-Walter Rix, and David W. Hogg. "The Milky Way Has No Distinct Thick Disk." *The Astrophysical Journal* 751.2 (2012): 131.

Buckley, Matthew R., and Patrick J. Fox. "Dark Matter Self-Interactions and Light Force Carriers." *Physical Review D* 81.8 (2010).

De Blok, W. J G. "The Core-Cusp Problem." *Advances in Astronomy* (2010).

Faber, S. M., and R. E. Jackson. "Velocity Dispersions and Mass-to-Light Ratios for Elliptical Galaxies." *The Astrophysical Journal* 204 (1976): 668–683.

"First Signs of Self-Interacting Dark Matter?" *ESO Press Release,* European Southern Observatory. http://www.eso.org/public/news/eso1514/

Goldberg, Haim, and Lawrence J. Hall. "A New Candidate for Dark Matter." *Physics Letters B* 174.2 (1986): 151–155.

Governato, F et al. "Bulgeless Dwarf Galaxies and Dark Matter Cores from Supernova-Driven Outflows." *Nature* 463.7278 (2010): 203–6.

Holmberg, Johan, and Chris Flynn. "The Local Surface Density of Disc Matter Mapped by Hipparcos." *Monthly Notices of the Royal Astronomical Society* 352.2 (2004): 440–446.

Kuijken, Konrad, and Gerard Gilmore. "The Galactic Disk Surface Mass Density and the Galactic Force K(z) at Z = 1.1 Kiloparsecs." *The Astrophysical Journal* 367 (1991): L9-L13.

Langdale, Jonathan. "Could There Be a Larger Dark World with Dark Interactions? There Is More Dark Matter than Visible." (2013). https://plus.google.com/+JonathanLangdale/posts/Es7M9VhiFNp.

Markevitch, M. et al. "Direct Constraints on the Dark Matter Self-Interaction Cross Section from the Merging Galaxy Cluster 1E 0657–56." *The Astrophysical Journal* 606.2 (2004): 819–824.

Moore, Ben et al. "Dark Matter Substructure within Galactic Halos." *The Astrophysical Journal* 524.1 (1999): L19–L22.

Oort, J. H. "The Force Exerted by the Stellar System in the Direction Perpendicular to the Galactic Plane and Some Related Problems." *Bulletin of the Astronomical Institutes of the Netherlands* 6 (1932): 249-287.

Oort, J. H. "Note on the determination of Kz and on the mass density near the Sun." *Bulletin of the Astronomical Institutes of the Netherlands* 15 (1960): 45.

Read, J I. "The Local Dark Matter Density." *Journal of Physics G: Nuclear and Particle Physics* 41.6 (2014): 063101.

Salucci, Paolo, and Annamaria Borriello. "The Intriguing Distribution of Dark Matter in Galaxies." *Particle Physics in the New Millennium* 616 (2003): 66–77.

Spergel, David N., and Paul J. Steinhardt. "Observational Evidence for Self-Interacting Cold Dark Matter." *Physical Review Letters* 84.17 (2000): 3760–3763.

Weinberg, David H. et al. "Cold Dark Matter: Controversies on Small Scales." *Proceedings of the National Academy of Sciences* (2015): http://arxiv.org/abs/1306.0913.

Weniger, Christoph. "A Tentative Gamma-Ray Line from Dark Matter Annihilation at the Fermi Large Area Telescope." *Journal of Cosmology and Astroparticle Physics* 2012.08 (2012).

Zhang, Lan, et al. "The gravitational potential near the sun from SEGUE K-dwarf kinematics." *The Astrophysical Journal* 772.2 (2013): 108.

CHAPTER 19

Cline, James M., Zuowei Liu, and Wei Xue. "Millicharged Atomic Dark Matter." *Physical Review D* 85.10 (2012): 101302.

Cooper, A. P. et al. "Galactic Stellar Haloes in the CDM Model." *Monthly Notices of the Royal Astronomical Society* 406.2 (2010): 744–766.

Dienes, Keith R., and Brooks Thomas. "Dynamical Dark Matter: A New Framework for Dark-Matter Physics." *Workshop on Dark Matter, Unification and Neutrino Physics: CETUP* 2012*. Vol. 1534. AIP Publishing, 2013. 57–77.

Fan, JiJi et al. "Dark-Disk Universe." *Physical Review Letters* 110.21 (2013): 211302.

Fan, JiJi et al. "Double-Disk Dark Matter." *Physics of the Dark Universe* 2.3 (2013): 139–156.

Foot, R. "Mirror Dark Matter: Cosmology, Galaxy Structure and Direct Detection." *International Journal of Modern Physics A* 29.11n12 (2014): 1430013.

Foot, R., H. Lew, and R.R. Volkas. "A Model with Fundamental Improper Spacetime Symmetries." *Physics Letters B* 272.1-2 (1991): 67–70.

Kaplan, David E et al. "Atomic Dark Matter." *Journal of Cosmology and Astroparticle Physics* 05 (2010). 21.

Kaplan, David E et al. "Dark Atoms: Asymmetry and Direct Detection." *Journal of Cosmology and Astroparticle Physics* 10 (2011): 19.

Pillepich, Annalisa et al. "The Distribution of Dark Matter in the Milky Way's Disk." *eprint arXiv:1308.1703* (2013).

Powell, Corey S. "Inside the Hunt for Dark Matter." *Popular Science*. (2013). http://www.popsci.com/article/science/inside-hunt-dark-matter.

Powell, Corey S. "The Possible Parallel Universe of Dark Matter." *Discover Magazine.com*. (2013). http://discovermagazine.com/2013/julyaug/21-the-possible-parallel-universe-of-dark-matter.

Purcell, Chris W., James S. Bullock, and Manoj Kaplinghat. "The Dark Disk of the Milky Way." *The Astrophysical Journal* 703.2 (2009): 2275–2284.

Read, J. I. et al. "Thin, Thick and Dark Discs in ËCDM." *Monthly Notices of the Royal Astronomical Society* 389.3 (2008): 1041–1057.

Rosen, Len. "Is There Only One Type of Dark Matter?" (2013). http://www.21stcentech.com/type-dark-matter/.

CHAPTER 20

Bovy, Jo, and Hans-Walter Rix. "A Direct Dynamical Measurement of the Milky Way's Disk Surface Density Profile, Disk Scale Length, and Dark Matter Profile at 4 Kpc $\lesssim$ R $\lesssim$ 9 Kpc." *The Astrophysical Journal* 779.2 (2013): 1–30.

Bovy, Jo, and Scott Tremaine. "On the Local Dark Matter Density." *The Astrophysical Journal* 756.1 (2012): 89.

Bruch, Tobias et al. "Dark Matter Disc Enhanced Neutrino Fluxes from the Sun and Earth." *Physics Letters B* 674.4-5 (2009): 250–256.

Bruch, Tobias et al. "Detecting the Milky Way's Dark Disk." *The Astrophysical Journal* 696.1 (2009): 920–923.

Buckley, Matthew R. et al. "Scattering, Damping, and Acoustic Oscillations: Simulating the Structure of Dark Matter Halos with Relativistic Force Carriers." *Physical Review D* 90.4 (2014): 043524.

Cartlidge, Edwin. "Do Dark-Matter Discs Envelop Galaxies?" *PhysicsWorld.com*. (2013). http://physicsworld.com/cws/article/news/2013/jun/03/do-dark-matter-discs-envelop-galaxies.

Cyr-Racine, Francis-Yan et al. "Constraints on Large-Scale Dark Acoustic Oscillations from Cosmology." *Physical Review D* 89.6 (2014).

Cyr-Racine, Francis-Yan, and Kris Sigurdson. "Cosmology of Atomic Dark Matter." *Physical Review D* 87.10 (2013).

Holmberg, J, and C Flynn. "The Local Density of Matter Mapped by Hipparcos." *Monthly Notices of the Royal Astronomical Society* 313.2 (2000): 209–216.

Kuijken, K., and G. Gilmore. "The Mass Distribution in the Galactic Disc - II - Determination of the Surface Mass Density of the Galactic Disc Near the Sun." *Monthly Notices of the Royal Astronomical Society* 239 (1989): 605–649.

Kuijken, Konrad, and Gerard Gilmore. "The Mass Distribution in the Galactic Disc. I - A Technique to Determine the Integral Surface Mass Density of the Disc near the Sun." *Monthly Notices of the Royal Astronomical Society* 239 (1989): 571–603.

March-Russell, John, Christopher McCabe, and Matthew McCullough. "Inelastic Dark Matter, Non-Standard Halos and the DAMA/LIBRA Results." *Journal of High Energy Physics* 2009.05 (2009).

McCullough, Matthew, and Lisa Randall. "Exothermic Double-Disk Dark Matter." *Journal of Cosmology and Astroparticle Physics* 2013.10 (2013): 58.

Motl, Luboš. "Exothermic Double-Disk Dark Matter." (2013). http://motls.blogspot.com/2013/07/exothermic-double-disk-dark-matter.html.

Nesti, Fabrizio, and Paolo Salucci. "The Dark Matter Halo of the Milky Way, AD 2013." *Journal of Cosmology and Astroparticle Physics* 2013.07 (2013): 16.

Randall, Lisa, and Jakub Scholtz. "Dissipative Dark Matter and the Andromeda Plane of Satellites." (2014). http://arxiv.org/abs/1412.1839.

Rix, Hans-Walter, and Jo Bovy. "The Milky Way's Stellar Disk." *The Astronomy and Astrophysics Review* 21.1 (2013): 61.

CHAPTER 21

Aron, Jacob. "Did Dark Matter Kill the Dinosaurs? Maybe. . . ." *New Scientist.* (2014). http://www.newscientist.com/article/dn25177-did-dark-matter-kill-the-dinosaurs-maybe.html#.VVYlfvlVhBc.

Choi, Charles Q. "Dark Matter Could Send Asteroids Crashing Into Earth: New Theory." (2014). http://www.space.com/25657-dark-matter-asteroid-impacts-earth-theory.html

Gibney, Elizabeth. "Did Dark Matter Kill the Dinosaurs?" *Nature.* (2014). http://www.nature.com/news/did-dark-matter-kill-the-dinosaurs-1.14839.

Nagai, Daisuke. "Viewpoint: Dark Matter May Play Role in Extinctions." *Physical Review Letters Physics* 7 (2014): 41.

Nair, Unni K. "Dinosaurs Extinction from Dark Matter?" (2014). http://guardianlv.com/2014/03/dinosaurs-extinction-from-dark-matter/.

Piggott, Mark. "Were Dinosaurs Killed by Disc of Dark Matter?" (2014). http://www.ibtimes.co.uk/were-dinosaurs-killed-by-disc-dark-matter-1439500.

Randall, Lisa, and Matthew Reece. "Dark Matter as a Trigger for Periodic Comet Impacts." *Physical Review Letters* 112.16 (2014): 161301.

Sharwood, Simon. "Dark Matter Killed the Dinosaurs, Boffins Suggest" *The Register.* 5 Mar. 2014. http://www.theregister.co.uk/2014/03/05/dark_matter_killed_the_dinosaurs_boffins_suggest/.

CONCLUSION

Bettencourt, Luís M A et al. "Growth, Innovation, Scaling, and the Pace of Life in Cities." *Proceedings of the National Academy of Sciences of the United States of America* 104.17 (2007): 7301–6.

Brynjolfsson, Erik, and Andrew McAfee. The Second Machine Age: Work, Progress, and Prosperity in a Time of Brilliant Technologies. W. W. Norton, 2014.

"Geoffrey West." (2015). http://www.santafe.edu/about/people/profile/Geoffrey%20West.

"On Care for Our Common Home." Encyclical Letter Laudato Si' of the Holy Father Francis (2015). http://w2.vatican.va/content/francesco/en/encyclicals/documents/papa-francesco_20150524_enciclica-laudato-si.html.

Weisman, Alan. *The World Without Us*. Reprint Edition. Picador, 2008.

West, Geoffrey. "Why Cities Keep Growing, Corporations and People Always Die, and Life Gets Faster." *The Edge*. (2011). http://edge.org/conversation/geoffrey-west.

INDEX

Italic page references indicate illustrations